De-convergence of Global Media Industries

Convergence has become a buzzword, referring on the one hand to the integration between computers, television, and mobile devices or between print, broadcast, and online media, and on the other hand, to the ownership of multiple content or distribution channels in media and communications. Yet while convergence among communication companies has been the major trend in the neoliberal era, the splintering of companies, de-convergence, is now gaining momentum in the communication market As the first comprehensive attempt to analyze the wave of de-convergence of the global media system in the context of globalization, this book makes sense of those transitions by looking at global trends and how global media firms have changed and developed their business paradigms from convergence to de-convergence. Dal Yong Jin traces the complex relationship between media industries, culture, and globalization by exploring it in a transitional yet contextually grounded framework, employing a political-economic analysis integrating empirical data analysis.

Dal Yong Jin is Associate Professor in the School of Communication at Simon Fraser University, Canada.

Routledge Research in Cultural and Media Studies

For a full list of titles in this series, please visit www.routledge.com

11 **Writers' Houses and the Making of Memory**
Edited by Harald Hendrix

12 **Autism and Representation**
Edited by Mark Osteen

13 **American Icons**
The Genesis of a National Visual Language
Benedikt Feldges

14 **The Practice of Public Art**
Edited by Cameron Cartiere and Shelly Willis

15 **Film and Television After DVD**
Edited by James Bennett and Tom Brown

16 **The Places and Spaces of Fashion, 1800–2007**
Edited by John Potvin

17 **Communicating in the Third Space**
Edited by Karin Ikas and Gerhard Wagner

18 **Deconstruction After 9/11**
Martin McQuillan

19 **The Contemporary Comic Book Superhero**
Edited by Angela Ndalianis

20 **Mobile Technologies**
From Telecommunications to Media
Edited by Gerard Goggin and Larissa Hjorth

21 **Dynamics and Performativity of Imagination**
The Image Between the Visible and the Invisible
Edited by Bernd Huppauf and Christoph Wulf

22 **Cities, Citizens, and Technologies**
Urban Life and Postmodernity
Paula Geyh

23 **Trauma and Media**
Theories, Histories, and Images
Allen Meek

24 **Letters, Postcards, Email**
Technologies of Presence
Esther Milne

25 **International Journalism and Democracy**
Civic Engagement Models from Around the World
Edited by Angela Romano

26 **Aesthetic Practices and Politics in Media, Music, and Art**
Performing Migration
Edited by Rocío G. Davis, Dorothea Fischer-Hornung, and Johanna C. Kardux

27 **Violence, Visual Culture, and the Black Male Body**
Cassandra Jackson

28 **Cognitive Poetics and Cultural Memory**
Russian Literary Mnemonics
Mikhail Gronas

29 Landscapes of Holocaust Postmemory
Brett Ashley Kaplan

30 Emotion, Genre, and Justice in Film and Television
E. Deidre Pribram

31 Audiobooks, Literature, and Sound Studies
Matthew Rubery

32 The Adaptation Industry
The Cultural Economy of Literary Adaptation
Simone Murray

33 Branding Post-Communist Nations
Marketizing National Identities in the "New" Europe
Edited by Nadia Kaneva

34 Science Fiction Film, Television, and Adaptation
Across the Screens
Edited by J. P. Telotte and Gerald Duchovnay

35 Art Platforms and Cultural Production on the Internet
Olga Goriunova

36 Queer Representation, Visibility, and Race in American Film and Television
Melanie E. S. Kohnen

37 Artificial Culture
Identity, Technology, and Bodies
Tama Leaver

38 Global Perspectives on Tarzan
From King of the Jungle to International Icon
Edited by Annette Wannamaker and Michelle Ann Abate

39 Studying Mobile Media
Cultural Technologies, Mobile Communication, and the iPhone
Edited by Larissa Hjorth, Jean Burgess, and Ingrid Richardson

40 Sport Beyond Television
The Internet, Digital Media and the Rise of Networked Media Sport
Brett Hutchins and David Rowe

41 Cultural Technologies
The Shaping of Culture in Media and Society
Edited by Göran Bolin

42 Violence and the Pornographic Imaginary
The Politics of Sex, Gender, and Aggression in Hardcore Pornography
Natalie Purcell

43 Ambiguities of Activism
Alter-Globalism and the Imperatives of Speed
Ingrid M. Hoofd

44 Generation X Goes Global
Mapping a Youth Culture in Motion
Christine Henseler

45 Forensic Science in Contemporary American Popular Culture
Gender, Crime, and Science
Lindsay Steenberg

46 Moral Panics, Social Fears, and the Media
Historical Perspectives
Edited by Siân Nicholas and Tom O'Malley

47 De-convergence of Global Media Industries
Dal Yong Jin

De-convergence of Global Media Industries

Dal Yong Jin

NEW YORK AND LONDON

First published 2013
by Routledge
711 Third Avenue, New York, NY 10017

Simultaneously published in the UK
by Routledge
2 Park Square, Milton Park, Abingdon, Oxfordshire OX14 4RN

First issued in paperback 2014

Routledge is an imprint of the Taylor and Francis Group, an informa business

Library of Congress Cataloging in Publication Data

Jin, Dal Yong, 1964–
De-convergence of global media industries / Dal Yong Jin.
pages cm. — (Routledge research in cultural and media studies ; 47)
Includes bibliographical references and index.
1. Mass media—Management. 2. Mass media and business. I. Title.
P96.M34J56 2013
302.23—dc23
2012039570

ISBN 978-0-415-62343-8 (hbk)
ISBN 978-1-138-92216-7 (pbk)
ISBN 978-0-203-58803-1 (ebk)

Typeset in Sabon
by Apex CoVantage, LLC

Contents

Figures

Tables

Preface

Convergence, referring not only to the consolidation of communication corporations but also to the integration of different technologies and culture, has radically reshaped the communication landscape over the past two decades. Global communication corporations, dealing both with media and telecommunications, have actively pursued convergence in order to control the whole media sector, from the production to the distribution of content through digital operating systems. Because of synergy effects, economies of scale, and the production and distribution of diverse cultural products, corporations not only in new media industries, dealing with the Internet, mobile, and cable, but also in old media and telecommunications industries, including newspaper, telephony, network broadcasting, and computers, have sought convergence. Media corporations believe that the synergized whole is worth more than the sum of its constituent parts, so they have expanded their scales in the digital era. The global communication system has indeed undergone a process of transformation under neoliberal globalization since the early 1980s when neoliberalism had just started as a major vehicle for change in the media and telecommunications industries.

The communication industries, however, have been structurally altered since the early 21st century, because major communication corporations have adopted de-convergence, which focuses on a few core businesses while splitting off/spinning off noncore sectors, in the midst of the failure of media convergence. Through two major economic crises in the early 21st century, from the dot-com bubble to the 2007–2008 financial downturn, global communication giants headquartered in the U.S., the U.K., France, and Canada have sold parts of their companies and split off and/or spun off their firms rather than pursuing consolidation. While convergence among communication companies has been the major trend in the neoliberal era, the de-convergence trend is gaining impetus.

In other words, over the last two decades, the global communication system has fundamentally changed. The neoliberal communication policy gained momentum in the U.S. and the U.K. in the 1980s and has become a driving force for the consolidation and reform of the telecom industries toward a market-oriented system around the world. Media and telecom companies, however, have begun to exhibit symptoms of what appears to be the same life-threatening disease in recent years. From Western countries to developing countries, overcapacity and severe competition among communication companies have led to bankruptcies and financial deficits. Consequently, many media companies in several countries have sought new survival strategies, such as spin-off and/or split-off strategies as well as counter-deregulation, which is de-convergence in the early 21st century.

As the first comprehensive attempt to analyze the wave of de-convergence of global media industries in the context of neoliberal globalization, I make sense of these transitions by looking at global trends and how global media firms have changed and developed their business paradigms from convergence to de-convergence. I engage with the complex relationship between media industries, culture, and globalization by exploring the changing nature of global media industries in a transitional yet contextually grounded framework. More specifically, I focus my attention on the change and continuity in the global communication system and culture by examining the wave of de-convergence, and I chronicle political, economic, technological, and cultural dimensions that have led to this dramatic change. I also map out how some of the key features of the global communication system have been reorganized and transnationalized since the early 1980s (i.e., how the transformation of the global communication system can be understood within the larger context of global political-economic shifts and accompanying technological development).

I use the time period between 1982 and 2012 because this time span includes three major historical events influencing the transformation of the communication industries in relation to neoliberal globalization. First, two countries, the U.K. and the U.S., began neoliberal transformation and globalization in the early 1980s; second, I also need to analyze the impact of the 1996 Telecommunications Act and the 1997 World Trade Organization (WTO) agreements on the transnationalization of the global communication industry in the 21st century. Finally, I deal with the current financial crisis, as a result of the failure of neoliberal globalization, which has greatly influenced the transformation of the global media industries in recent years.

As a starting point of discussion, I first examine the history and practice of convergence and de-convergence by examining mergers and acquisitions (M&As) in the communication sector between 1982 and 2012. I examine the role of transnational corporations (TNCs), mainly addressing how they were involved in the reshaping of the global telecom system. I discuss the relationship between national policies and institutions, on the one hand, and international organizations, including the International Telecommunication Union (ITU) and the WTO, on the other. Then, I attempt to clarify whether the transformation of the global media system has influenced the crisis in the communication sector in recent years, and I thereafter articulate how many telecom companies around the world are trying to find new strategies and/or new business models, differentiating them from neoliberal telecom restructuring. In addition, I discuss why and how communication companies have pursued de-convergence by employing split-off and spin-off strategies as their business models after and/or with the failure of the mergers. I investigate the macro-level trend of changing corporate strategies within the communication industries based on whether they have achieved synergy effects, such as increasing profits, revenues, and stock prices, or whether they have not. This means that the primary focus of the book is the contextualization of the changing corporate business trend by historically analyzing the rise and fall of media convergence, and thereafter media de-convergence. These discussions lead to the articulation of whether de-convergence could become a solid new trend replacing convergence, or whether convergence will regain its power in dominating the communication industry. Throughout the discussion, I hope to shed light on current developments and place them in a perspective that has relevance for future communication policy/industry directions.

Portions of Chapters 5 and 8 originally appeared as materials in the following journals: "The New-wave of De-convergence: A New Business Model of the Communication Industry in the 21st Century," *Media, Culture and Society 34*(6): 761–772 (2012); "Transforming the Global Film Industries: Horizontal Integration and Vertical Concentration Amid Neoliberal Globalization," *International Communication Gazette 74*(5): 405–422 (2012); and "De-convergence: A Shifting Business Trend in the U.S. Digital Media Industries," *Journal of Media Economics and Culture 7*(1): 3–44 (2009).

1 Introduction

The communication industries have been central to the development of contemporary societies, and the primary roles of the communication industries have been crucial to the establishment of nation-states and the development of the public sphere and democracy, as well as the rise of capitalist commercial enterprises. Due to their pivotal roles in association with human freedom and national identity, as well as in the formation of political culture, many governments in the world have utilized their media/telecommunications policy as the principle means by which people are informed and entertained (Flew, 2007).

Since the early 1980s, the global communication industries have undergone a process of structural transformation in order to effectively produce and distribute their cultural products and services. Instead of emphasizing characteristics of the public sphere, the communication industries have expanded geographically in the midst of newly introduced neoliberal globalization. Global communication corporations, dealing with both media (e.g., Viacom, Time Warner, and Vivendi) and telecommunications (e.g., AT&T, Comcast, NTT, and Bell Canada Enterprise), have actively pursued convergence as a form of mergers and acquisitions (M&As) in order to control the whole media sector, from the production to the distribution of content through digital operating systems. Corporations in not only new media industries, including the Internet, social media (such as YouTube, Facebook, and Twitter), mobile, and cable, but also old media and telecommunications industries, such as newspaper, telephony, network broadcasting, and computers, have sought convergence because of synergy effects, economies of scale, and the production and distribution of diverse cultural products. Communication corporations calculate that the synergized whole is worth more than the sum of its constituent parts (Murray, 2003), so through vertical and horizontal integration, they have expanded their scale in the digital era.

The communication industries have been structurally altered since the early 21st century because major communication corporations have adopted de-convergence in the midst of the failure of media convergence. While M&As among communication firms have been the major trend in the neoliberal era, mega transactions in the communication industries have not achieved intended/anticipated synergy effects, and many communication corporations are now focusing on a few core businesses while splitting off/spinning off noncore sectors, which are now gaining momentum in the global market (Jin, 2009a; 2011a). Global communication giants headquartered in the U.S., the U.K., France, and Canada have begun to sell parts of their companies and split off and/or spin off their firms rather than consolidating their corporations. Several mega communication giants, including CBS-Viacom, NBC Universal, Vivendi, and AT&T have experienced severe setbacks after their mergers, and they have strategically utilized de-convergence, resulting in the growth of the communication sector (Edge, 2010; Thornton and Keith, 2009; Albarran and Gormly, 2004; Adams, 2002). As the first comprehensive attempt to analyze the wave of de-convergence within the global communication system in the context of neoliberal globalization, in addition to significant convergence phenomena, this book makes sense of those transitions by looking at global trends and how global communication firms have changed and developed their business paradigms from convergence to de-convergence. It engages with the complex relationship between communication industries, culture, and globalization by exploring it in a transitional yet contextually grounded framework.

NEOLIBERALISM, GLOBALIZATION, AND FINANCIALIZATION IN THE COMMUNICATION INDUSTRIES

Changes in capitalism over the last several decades have been commonly characterized using a trio of terms: neoliberalism, globalization, and financialization (Foster, 2007). These three terms are not mutually exclusive, and sometimes they are used interchangeably in understanding contemporary capitalism and society (Jin, 2010). The global communication industries between the early 1980s and early 21st century have been an exemplary case of neoliberal transformation theory. The restructuring of the communication sector has been conducted under the banner of deregulation and liberalization not only in Western countries where this trend is most pronounced, but around the world, including the countries of the old socialist bloc as well as Asian and Latin American countries (McChesney, 2008; 1999). Several communication scholars, including Dan Schiller (1999b; 2007), Robert McChesney (2008), and Terry Flew (2007) point out that the global communication system has become increasingly transnational with the rise of neoliberalism, referring to the policies that maximize the role of markets and profit making and minimize the role of nonmarket institutions through deregulation and privatization (Friedman, 1982). Across the world, the vast majority of governments have introduced economic liberalization measures, including a reduction in government intervention in product markets, opening the domestic markets and privatizing the financial and public sectors, such as telecommunications and broadcasting industries. Neoliberalism drives the restructuring of national economies and culture, and this has consequences for communication industries. The communication industries in Asia and Latin America, as well as in the U.S., have consequently undergone a substantial transformation commonly referred to as neoliberal reform.

More specifically, over the last two decades, global communication industries have experienced dramatic neoliberal transnationalization. Critical political economists argue that neoliberalism has expanded and enhanced corporate profit making opportunities (Hart-Landsberg, 2006). Several theoreticians (McChesney, 2008; Flew, 2007; Bagdikian, 2004) also emphasize that neoliberalism has resulted in the concentration of ownership within a few mega media giants. As Robert McChesney argues (2001, p. 2), neoliberalism unleashed national and international politics maximally supportive of business domination of all social affairs. As he correctly points out, "the centerpiece of neoliberal policies is invariably a call for commercial communication markets to be deregulated." As a result of deregulation, privatization, and liberalization in the communication sector and other economic sectors, for example, broadcasting industries have begun to continuously transform themselves into the market-oriented communication system (McChesney, 2008; Jin, 2007). The aggressive promotion of the benefits of neoliberal reform by the U.S. and the U.K. massively increases both the operating space available to private corporations and the resources at their disposal (Murdock, 2006). Furthermore, Grant and Wood argue (2004, p. 85) that the global marketplaces of publishing, recorded music, film production, and broadcasting have all come under the control of giant corporations, mainly U.S. owned or U.S. based. As these scholars emphasize, neoliberalism engineers the restructuring of national economies and boundaries, and this has consequences for communication industries.

While the trend of acquiring multimedia and multifunctional networks has expedited the integration process (Schiller, 2007), the high level of M&As in the Internet services industries has also been expanded with the rapid employment of neoliberal globalization—the integration of the global economy into the liberal capitalist market economy controlled by a few Western countries. Neoliberal globalization is characterized by interlocking features, including policies that promote liberalization, deregulation, privatization, and capital investment (Lindio-McGovern, 2007). Neoliberal globalization began in the early 1980s and continued throughout the 1990s and the early 21st century, although corporate integration started several decades ago. The liberalization of controls on foreign direct investment and movements of capital has fostered vertical and horizontal integration through M&As (Kelsey, 2006, p. 196). The Telecommunications Act of 1996 in the U.S. and the 1997 WTO agreement have especially expedited mega M&As in the communication industries, because these two historical events alongside transnational corporations (TNCs)—whose policies and practices serve the interests of monopoly capital—are the major instruments of neoliberal globalization (Lindio-McGovern, 2007, p. 2), and the Internet services industry has been one of the most significant sectors in the market (Chambers & Howard, 2005; Geradin & Luff, 2004; Chon et al., 2003). The communication industries have taken full advantage of worldwide deregulatory and privatization trends to make ever larger conglomerates mainly through M&As (Gershon, 2006).

Neoliberal globalization is also closely connected to financialization, which is one of the most significant capital accumulation strategies in the contemporary capitalist system. The term financialization has varying definitions, and in its broader concept, financial sectors have come to play a more dominant part relative to the economy as a whole (Davis, 2011, p. 242). Gerald Epstein (2005, p. 3) defines financialization as "the increasing role of financial motives, financial markets, financial actors, and financial institutions in the operation of the domestic and international economics." Krippner (2005, p. 174) also points out, "financialization refers to a pattern of capital accumulation in which profits accrue primarily through financial channels rather than through trade and commodity production." Meanwhile, Foster (2007) succinctly claims the financialization of capitalism as the shift in gravity of economic activity from production (and even from much of the growing service sector) to finance, which is thus one of the key issues of our time. While various authors use the term financialization differently, which means that scholars in different fields define financialization as a multifaceted, general tendency of capitalism in our time (Erturk et al., 2008), and they also commonly use it to describe the increasing globalization of financial markets. Since financialization refers to a pattern of capital accumulation in which profits accrue primarily through financial channels rather than through trade and commodity production, we need to primarily analyze the convergence of non-Western and Western media corporations as a form of M&As as a major part of financialization in the communication industries (Jin, 2010).

There are only a few historical and empirical references to the rapidly changing global communication system through capital investment. Several scholarly papers (Jung, 2004; Peltier, 2004; Chan-Olmsted, 1998; Howard, 1998) have focused on the issues of M&As; however, they have mainly explored the economic effects of M&As and strategies for M&As—meaning these previous works emphasized the structural changes of the global communication system not from political economy perspectives but rather from media management perspectives. They are also not comprehensive because of the limited time period examined and the shortage of industries in the communication sector studied.[1] From the political economy approach, though, Dwayne Winseck (2011, 143) points out, "the logic of financialization is particularly important to recent developments across the media industries because it has, paradoxically, not only created greater media concentration but also bloated media giants that have sometimes stumbled badly and occasionally been brought to their knees by the two global financial crises of the 21st century," with the case of the Canadian media industry. Through ongoing discussion, this book sheds light on current debates on the neoliberal transformation in the global communication system and its consequential result of de-convergence, which eventually contributes to the development of current theories of neoliberal globalization and financialization.

GROWING CONVERGENCE AMID NEOLIBERAL GLOBALIZATION

Convergence has been a buzzword and the notion of convergence has grown in popularity since the 1990s, although it has a longer history. Convergence is as old as the telegraph, and the promises and challenges we associated with the Internet were anticipated by that mid-19th-century technology (Winseck & Pike, 2007). Rich Gordon (2003) also points out that the concept of convergence comes originally from the world of science and mathematics, and only later to political science. In the communication sector, the development of computers and networks in the 1960s and 1970s facilitated the usage of convergence. In the late 1990s and the early 21st century, convergence even further becomes one of the central developments taking place across the media, telecommunications, and information sectors of the communication industry (Mosco & McKercher, 2006).

Due to its diverse origins during a longer historical background, media convergence is not simple and has multiple meanings. For some scholars, convergence is synonymous with media consolidation. For others, it describes what happens when a new, multifunction device, such as a telephone/cable/Internet modem, does jobs that previously would have required two or more appliances (Schnaars et al., 2008, cited in Thornton and Keith, 2009). Most important, to converge means to come together. In the context of communication technologies, one talks about the coming together of several discrete technologies to create a hybrid technology (Mosco & McKercher, 2006):

> Technological convergence typically means the integration of the devices that these industries use as well as the information they process, distribute, and exchange over and through these devices. By integrating computers and telecommunications, the Internet is now an iconic example of technological convergence. This form of convergence is linked to, and partly responsible for, the convergence of once separated industries into a common arena providing electronic information and communication services. (Mosco & McKercher, 2006, p. 734)

Another important element is a commercial convergence between media content companies, such as television and radio networks and transmission channels (for example, those dealing with telephone cable networks). Likewise, convergence in communication has diverse meanings.

More specifically, media convergence can be categorized in three major areas: the flow of content across multiple media platforms; the cooperation between multiple media industries; and the migratory behavior of media audiences who will go almost anywhere in search of the kinds of entertainment experience they want (Jenkins, 2006, pp. 2–3). Other scholars (Baldwin et al., 1996; Wirtz, 2001) also view media convergence from three different perspectives: consolidation through industry alliances and mergers; the combination of technology and network platforms; and the integration between services and markets. Meanwhile, Gordon (2003) points out that the term convergence has been applied to corporate strategies, to technological developments, to marketing efforts, and to job descriptions. For Gordon, convergence especially means the ownership of multiple content or distribution channels.

These concepts are not mutually exclusive due to the close relationship between media structure and content, and, as it is commonly agreed, convergence cannot be done without the integration of production between the old and new media (Jin, 2011a). Through vertical and horizontal integration, a company like Time Warner controls interests in several different areas, including film, television, books, games, the Web, music, and countless other sectors. The result has been the restructuring of cultural production around multiple forms of presumed synergies (Jenkins, 2001). Indeed, starting in the early 1990s, the wave of convergence has witnessed increasing concentration of ownership of mainstream media properties in the hands of a select group of transnational media corporations, a phenomenon achieved both through outright acquisition and vast mergers (Murray, 2003). The prime goal of corporations is to grow profitably and that, at least, is the logic assumed and asserted by the 'rational markets' discourses which have underpinned merger activity in recent times; in practice, however, such idealizations may often be tainted or overridden by short-term rent seeking, financial engineering, or self-interest and the like on the part of top level management and their professional financial and legal advisors.

Communication corporations pursue convergence due in large part to the maximization of profits through the concentration of media companies, because vertical and horizontal integration are expected to bring synergy effects (Noam, 2009; Chambers & Howard, 2005; Chan-Olmsted, 1998). Communication companies have merged and acquired one another in the name of convergence and the supposed synergies that would result. Convergence therefore refers to both the integration of different technologies and the integration of industries across different business sectors and different industries. While some corporations have horizontally integrated within the same areas (for example, between traditional network broadcasting and new cable broadcasting companies and/or between the Internet and telephony companies),[2] other media companies have vertically integrated (for example, between broadcasting and film industries and/or broadly between telecommunications and broadcasting industries).[3] While convergence denotes the technological integration that powers new media technologies, as well as distinctive new media forms and formats (Jenkins, 2006), it also means the integration of big companies that make use of new media. In essence, interconnected technologies and large integrated companies create the convergence it takes to make a revolution (Mosco, 2008).

Convergence is particularly a challenge for many traditional media companies that attempt to venture into the Internet market because current media convergence has been fuelled by increasingly pervasive digital technologies and neoliberal communication policies (Huang & Heider, 2008, p. 105; Bar & Sandvig, 2008). For those in the old media industries, such as network broadcasters and newspaper companies, convergence has become a great opportunity to integrate with the new media sector, and such industrial convergence—the new trend of acquiring multimedia and multifunctional networks—is the integration of (digital) hardware and (content) software (Noll, 2003). This new trend of acquiring multimedia and multifunctional networks has facilitated the convergence process (Schiller, 2007). Again, since the 1980s, the majority of governments, not only in the U.S. but also in non-Western countries, have introduced political and economic liberalization measures (Friedman, 2002). The aggressive promotion of the benefits of neoliberal reform by the U.S. and the U.K. massively increases the resources at their disposal through the privatization of telecommunications and broadcasting companies in many countries, which has changed the map of the communication industry (Murdock, 2006).[4]

Before the introduction of and their integration with neoliberal globalization, the majority of media companies had focused on their own core business areas, partially because government policies, including antitrust laws and cross-ownership restraints, sought to define them distinctly and kept them separate (Baldwin et el., 1996). Indeed, the dangers of ownership concentration in the communication industry were addressed by a combination of antitrust and regulatory policies that attempted to attenuate the amalgamation of corporate power. Corporate mergers and the consolidation of ownership in the communication arena have long been sources of concern because they potentially hurt a diversity of voices. U.S. regulatory and antitrust policy has thus traditionally attempted to address the potential negative implications of concentrated media power by securing the diversity of voices structurally, largely through rules regarding ownership (Horwitz, 2005, pp. 181–182). Therefore, Time Inc. was known as a publication company, and Viacom was famous for its TV syndication and cable businesses, whereas News Corporation was considered a newspaper company based in Australia.

However, probusiness neoliberal communication policies have lifted the barriers of new investment opportunities, resulting in the concentration of ownership through media convergence in the hands of a few media giants (McChesney, 2008). Radical changes in ownership and control through convergence were the necessary foundations for the massive enlargement of corporate capital investment in networks and for the spectacular growth of transnational business that pyramided on top of it (Chakravartty & Schiller, 2010). Vertical and horizontal integration secures corporations' access to new markets and new sectors of the communication industries, although the consequence is the oligopolistic control of the world communication market by a few giant corporations (Jin, 2007). In other words, changing media policies have facilitated the concentration of media ownership. The privatization of public broadcasting and telecommunications and the liberalization of domestic markets have blurred the distinction between the old media and the new media, and media companies have become multimedia companies through M&As in the midst of neoliberal reform (McPhail, 2006). Deregulation and, in particular, the loosening of regulations in cross-ownership among communication industries has become one of the most significant factors in expanding their business areas beyond the traditional core business realms in many media companies (Thussu, 2006). Thus, media corporations have become larger and presumably more powerful, and ownership regulations have been rescinded or struck down over the last two decades (Horwitz, 2005).

THE EMERGENCE OF DE-CONVERGENCE IN THE COMMUNICATION INDUSTRIES

Media convergence, which has swiftly grown based on neoliberal policies and digitization, has been controversial because it has often failed in producing promised synergy effects. Corporations that had acquired too many companies in too many unrelated lines of business in the conglomerate movement became impossible to manage strategically and began to unravel (Lazonick, 2009). As a result, de-convergence, which is the breakup of companies in various ways, has gradually replaced convergence in the communication industries. The de-convergence trend is not new, of course. As Manuel Castells already pointed out,[5]

> throughout the 1990s, media tycoons pursued the dream of convergence between computers, the Internet, and the media. The key word was multimedia, and its materialization was the magic box that would sit in our living room and could, at our command, open a global window to endless possibilities of interactive communication in video, audio, and text format. Yet, the business experiments on media convergence carried on since the early 1990s have ended in failure. Most of the forms of convergence did not make money. Indeed, traditional media companies are not generating any profits from their Internet ventures. And the prospects are unlikely to change in the near future. (2001, pp. 188–190)

In the mid-1990s, a few media scholars documented the progress of disintegration in the communication sector (Gilder, 1994, cited in Mueller, 1999).

With two major cases of mergers failing (AOL-Time Warner and Vivendi Universal), Albarran and Gormly (2004) argue that convergence has given way to divergence or de-merger. As Albarran and Gormly (2004) correctly argue, fewer than half of the mergers survived, dating back to the inception of the merger process. Stephanie Peltier (2004, p. 271) also points out, "it became obvious that the attempt to build media giants through M&As undoubtedly failed, because many media firms after M&As did not improve economic performance measured by profit margins." After their empirical study of the U.S. newspaper industry, Thornton and Keith (2009) claimed that the convergence of broadcast-print has broadly failed and convergence is an outdated form. This implies that the print media industry, as independent mass media instead of converged media, continues to publish in hard copy, although they publish digital formats. Marc Edge (2010) also addresses that de-convergence in Canada has been seen over the past several years as a reaction to the well-documented business failures of convergence.

Meanwhile, it is very important to understand that the changing behaviors of audiences have made differences (Jenkins, 2006), because they are some of the most significant elements for both convergence and de-convergence—although this book does not intend to raise the question of whether media audiences and technology users have been homogenized in the midst of media convergence, or whether they still enjoy distinctive cultural values in different countries. As is well chronicled, one aspect of globalization is the convergence of income, media, and technology, and people expect convergence to lead to homogeneous consumer needs, tastes, and lifestyles (De Mooij, 2003, p. 183). Media corporations believe that they benefit from homogenous consumer styles backed by advertising through cross-border media channels, which creates convergence mania. In opposition to their assumptions, however, several large multinational media firms have seen their profits decline because centralized control lacks local sensitivity, and these firms are consequently changing their strategies from global to local. Some of the myths surrounding global communication corporations are convergence of consumer behavior, the existence of universal values, and global communities with similar values; however, consumers in many countries tend to diverge with respect to how people use media products (De Mooij, 2003). This implies that changing behaviors of audiences, which are major elements for convergence, have ironically become some of the most significant factors for media de-convergence.

The major de-convergence process mainly started in the early 21st century, as some scholars acknowledged the failure of convergence. De-convergence is an emerging trend in the communication industry, although discussions have not been supported widely yet, nor discussed in 'full swing,' as a reflection of time limit. By its own nature, there is no clear definition of de-convergence, because mega media giants, such as Time Warner and Viacom have only begun to find a new business model after their losses in revenues and profits with convergence.[6]

In this book, de-convergence primarily refers to the business activities in which communication companies strategically decrease the magnitude of firms in order to regain their profits and public images and/or to survive in the market with several business strategies. As we define convergence as the consolidation of firms within the industry influenced by digitization and probusiness government policies, we designate de-convergence as the form of dis-integration of corporations by either selling parts of the shares to other companies or splitting off and/or spinning off of their companies.[7] De-consolidated companies through spin-offs and/or spilt-offs are operating as independent corporations separated from a parent corporation in the market; therefore, they are not part of the parent company anymore, although the parent company still holds some shares of these companies in many cases.

While there are several significant implications, de-convergence is especially important, primarily because this process has changed the ownership structure that was concentrated through convergence. As Gordon (2003) points out, at the highest level of today's media conglomerates, convergence means the ownership of multiple content or distribution channels. Several media critics (Bagdikian, 2002; McChesney, 1999) worried that the concentration of ownership and control of content by a handful of companies with an interest in preserving the status quo would stifle the diversity of voices necessary to produce an accurate picture of reality in news coverage and media culture. Therefore, it is fundamental to raise the question of whether de-convergence has resolved these significant policy issues, including diversity for democracy, or whether it has reflected only the commercial interests of owners of communication corporations who desire more benefits through the de-convergence process.

PURPOSES OF THE BOOK

This book focuses on change and continuity in the global communication system and culture by analyzing the wave of de-convergence, as well as convergence, and it chronicles political, economic, technological, and cultural dimensions that have led to this dramatic change in the global communication system, which includes broadcasting, advertising, movie, the press, the Internet, game, and telecommunications industries. It maps out how some of the key features of the global communication system have been reorganized and transnationalized since the early1980s (i.e., how the transformation of the global communication system can be understood within the larger context of global political-economic shifts and accompanying technological development).

There are three major goals in relation to industry, people, and globalization. As a starting point of discussion, this book first examines the history and practice of convergence and de-convergence by examining

M&As in the communication sector between 1982 and 2012. It investigates how the restructuring of the communication industries can be understood within the larger context of neoliberal communication policy changes, such as deregulation and liberalization measures, and accompanying technological development, particularly digital technologies. Second, it discusses why and how communication companies have pursued de-convergence by employing split-off and spin-off strategies as their business models after the failure of the mergers. As the AOL-Time Warner and CBS-Viacom cases prove, several major merged media corporations have been separated into several smaller media corporations. This book maps out the macrolevel trend of changing corporate strategies within the communication industries based on whether they have achieved synergy effects, such as increasing profits, revenues, and stock prices, or not, instead of attempting to analyze microeconomic reasons for the failure of individual M&As. This means that the primary focus of the book is the contextualization of this changing corporate business trend by historically analyzing the rise and fall of media convergence, and thereafter the emerging corporate paradigm of media de-convergence. These discussions lead to the articulation of whether de-convergence could become a solid new trend replacing convergence, or whether convergence will regain its power in dominating the communication industry. Media convergence, occurring at various intersections of media technologies and industries, is not an end state but an ongoing process. However, with the failure of neoliberal globalization in many areas, de-convergence has rapidly become a strong trend in the communication industries. Therefore, it is crucial to understand the interplay of convergence and de-convergence from political economic perspectives.

This book uses the time period between 1982 and 2012 because this time span includes three major historical events influencing the transformation of the communication industries. First, two countries, the U.K. and the U.S., began the neoliberal transformation and globalization in the early 1980s. Second, we also need to analyze the impact of the 1996 Telecommunications Act in the U.S. and the 1997 WTO agreements on the transnationalization of the global communication industry in the 21st century. (While there are several significant government policies, such as the Broadcasting Act of 1996 and the Telecommunication Act of 1993 in Canada, the major focus of the book is global in scope, so we focus on the global trend with an analysis of the impact of the two major deregulation measures on the global communication industries.) Finally, the book deals with the current financial crisis in the early 21st century as a result of the failure of neoliberal globalization, which has greatly influenced the transformation of the global communication industries in recent years. This book eventually determines whether the global communication industries have changed their corporate strategies, from convergence to de-convergence, during periods of economic crises.

METHODS OF THE STUDY

This book examines how the transformation of the global communication system can be understood within the larger context of global political-economic shifts and accompanying technological development. For this, it analyzes two major resources by means of historical and political economy approaches. Historically, the first major source of data used in this study is the SDC Platinum database compiled by Thomson Financial Company, which includes all corporate transactions, private as well as public, involving at least 5% of ownership of a company between 1982 and 2009. The second major resource that I analyze is the 'M&A Almanac' in *Mergers and Acquisitions*, which is the dealmakers' trade journal that Thompson Financial Company published every month between 1992 and 2010 in order to discern the solid trend of M&As in the international communication sector. Meanwhile, government documents from several countries, corporate annual reports, and data from international agencies such as UNESCO, the OECD (Organisation for Economic Co-operation and Development), and the ITU (International Telecommunication Union) are useful as secondary materials to support the discussions. The political economy approach is crucial in the analysis of the communication industries, because it examines the nature of the relationship between media and communication systems and the broader social structure of society, mostly of global capitalism (Garnham, 2011; McChesney, 2008, Schiller, 2007). As Vincent Mosco (2009, p. 2) argues, the political economy of communication is "the study of the social relations that mutually constitute the production, distribution, and consumption of communication resources, such as newspapers, videos, films, and audiences." Given that information and communication technologies (ICTs) drive the acceleration of global capitalism (Hope, 2011, p. 524), the political economy approach provides a fundamental tool to analyze the critical relationship between the communication industries, which are highly capitalized, and the modern capitalism, which is itself being global.

ORGANIZATION OF THE BOOK

In Chapter 2, I deal with how the global communication industries have changed from the early 1980s as a result of adopting neoliberal policies. As a comprehensive overview of all the other chapters, this chapter examines the neoliberal transnationalization of the communication industries via consolidation. It discusses the way in which the two most significant events in the global communication system—the 1996 Telecommunications Act and the 1997 WTO agreements—have changed the landscape of the global communication industry. Before discussing each media industry, this chapter comparatively analyzes the major differences and similarities in the deal market so that it provides fundamental guidelines for deeply understanding different industries in several of the following chapters.

Chapter 3 explores the transformation of the broadcasting industry. It maps out how some of the key features of the global TV system have been reorganized since the mid-1980s—that is, how the transformation of the global TV industry system can be understood within the larger context of global political-economic shifts. Specifically, I explore the changing structure of the broadcasting industry by examining consolidation. That is, I analyze foreign and domestic investment activities of the TV industry through M&As in the past 20 years. Then I discuss the role of national governments and domestic television industries in the transformation of the television system. In Chapter 4, I examine the neoliberal transnationalization of the advertising industry via convergence. This chapter explores the role of the U.S. corporations, which have been considered the key players in the global advertising system, in the global M&A market to determine whether non-Western countries have expanded their influence with their capital on the global advertising system.

Chapter 5 intends to examine the historical development of the global film industries primarily though horizontal integration between the late 20th and the early 21st century. Its essence lies in an empirical analysis of the structural change and dynamic of the film industries. It explores the role of U.S. film corporations—considered the key players in the global film market through Hollywood movies—to determine whether the U.S. has taken a pivotal role in the global M&A market, as in the case of the cultural market. This leads me to raise the fundamental question of whether film corporations in non-Western countries have expanded their influence on the global market so that they can diminish an asymmetrical power relation between the West and the East.

In Chapter 6, I study the role of TNCs, mainly examining how they were involved in the reshaping of the global telecom system. I discuss the relationship between national policies and institutions, on the one hand, and international organizations, including the International Telecommunication Union (ITU) and the WTO, on the other hand. Then I attempt to clarify whether the transformation of the global telecom system has influenced the telecom crisis in recent years, and I thereafter articulate how many telecom companies around the world are trying to find new strategies and/or new business models, differentiating them from neoliberal telecom restructuring. Through the discussion, I hope to shed light on current developments and place them in a perspective that has relevance for future telecom policy directions.

Chapter 7 investigates how the transformation of the global Internet services industries can be understood within the larger context of global political-economic shifts. It examines what conditions set the agenda for media and telecom corporations in order to analyze a long-term transformation in the global Internet services system. It also discusses the role of TNCs mainly by studying how they are involved in the reshaping of the global Internet services industries. It later calibrates why and how communication giants in Western countries, particularly the U.S. Internet services industries, have pursued disintegration, while it discusses whether non-Western countries have increased their capital power in the midst of the shifting media ecology in the 21st century.

In Chapter 8, I map out the history and practice of de-convergence by examining M&As in the communication sector. This chapter analyzes why many mega media deals have failed instead of achieving promised synergies by investigating 103 mega communication M&As occurring between 1998 and 2007. The chapter discusses why and how communication companies have pursued de-convergence by employing split-off and spin-off strategies as their business models after the failure of the mergers. Again, this book will not attempt to analyze microeconomic reasons for the failure of individual M&As. Instead, it discusses the macrolevel trend of changing corporate strategies within the communication industries based on whether they have achieved synergy effects, such as increasing profits, revenues, and stock prices, or not. This means that the primary focus of the chapter and of the book is the contextualization of the changing corporate business trend by historically analyzing the rise and fall of media convergence, and thereafter media de-convergence.

Chapter 9 examines both convergence and de-convergence in news and journalism through the analysis of the newspaper industry. It determines whether the original print-broadcast partnership model of convergence is being sustained, and it analyzes why and how the print media industry in Western countries has pursued convergence and later de-convergence, while discussing whether non-Western countries have increased their capital power in the global newspaper market, which has been considered as a local and/or national market instead of the global market in nature. Chapter 10 includes summaries of each chapter and provides some theoretical considerations for further studies.

Part I

Convergence of the Global Media Industries

2 Media Convergence of the Global Media Industries

The global communication industries have undergone a process of neoliberal transnationalization over the last three decades. As communication industries have expanded geographically through foreign direct investment (FDI) and acquisition, they have reorganized their structures in order to effectively produce and distribute their cultural products and services. From broadcasting to advertising agencies, and from newspaper companies to telecommunications firms, communication industries have sought to mediate increasing globalization of the world economy and culture through M&As. Only 20 years ago, the communication industries achieved limited integration reflecting their relatively small stake in the global industries. However, communication industries have substantially expanded their roles through capital investment, and they have become some of the largest industries in the global deal markets as a result of vertical and horizontal integration. There has been a trend toward the neoliberal globalization of communication services since the mid-1980s. During the 1980s, the politics of neoliberal communication reform took hold in dozens of nations following the U.S. and U.K.

This chapter, as a comprehensive overview of the structural change and a comparative analysis of the global communication industries, examines the neoliberal transnationalization of the communication industries via corporate convergence. It analyzes the foreign and domestic investment activities of the communication industries through M&As in broadcast, advertising, newspaper, and telecommunications (basic services) industries between 1982 and 2012. The chapter discusses the way in which the two most significant events in the global communication system—the 1996 Telecommunications Act and the 1997 WTO agreements—have changed the landscape of the global communication industry as major neoliberal reforms. Finally, it explores the role of U.S. corporations, which have been considered the key players in the global communication system in the global M&A market, to determine whether non-Western countries have expanded their influence with their capital within the global communication system in the midst of neoliberal globalization.

RISE AND FALL OF THE GLOBAL COMMUNICATION INDUSTRIES

Since the introduction of neoliberal economic and cultural policies in many countries, there has been a dramatic restructuring of the global media industries. Media industries have become major components in national economies, and the emergence of the global commercial media market has become real. The newly developing global media system has been dominated by a handful of transnational media corporations. In addition to the centralization of media power, the major feature of the global media order is its thoroughgoing commercialism and an associated marked decline in the relative importance of public broadcasting and the applicability of public service standards (Herman & McChesney, 1997, p. 1).

Most of all, the emergence of the communication industries as part of the major global economy has been discernible. The global communication industries were previously considered as a marginal segment of the global economy; however, they are considerably growing and becoming a large part of the national economy, as well as the global economy. According to the *Global Entertainment and Media Outlook: 2000–2004*, which analyzed 10 industry sectors within world media, spending of the U.S., Europe, and Asia/Pacific was projected at $913 billion in 2003, up from $689 billion in 1999.[1] Among these, the U.S., as the largest media market, accounted for 45.5% with $416 billion in spending, followed by Japan with $68.2 billion and the U.K. with $41 billion. Other Asian and Pacific countries' spending was as follows: China ($18.7 billion), Australia ($10.3 billion), Korea ($9.3 billion), Hong Kong ($2 billion), Indonesia ($1.7 billion), and New Zealand ($1.5 billion) (PricewaterhouseCoopers, 2000).[2]

The emergence of the communication industries in the early 21st century has been more phenomenal due to the rapid growth of the Internet industry. As Table 2.1 shows, the overall global market size has jumped by 23.2%, from $863.9 billion in 2004 to $1,064.5 billion in 2010 (estimation). While the current data differs from that mentioned above, it still proves the rapid growth of the communication industries in the early 21st century. Among the industries shown in Table 2.1, the Internet has become the largest industry since 2007. Until then, TV subscriptions and licensee fees and TV advertising were much bigger than the Internet industry; however, as a reflection of its growing role, the Internet industry has become a nondisputable giant industry in recent years. In fact, the Internet market increased 116% between 2004 and 2010 while music, magazine, and newspaper industries showed a downturn. If Internet advertising and video games were to be included as part of the Internet market, the massive role of the Internet industry would be even bigger than this figure.

Interestingly enough, the rapid growth of the Internet sector and the concentration of the Internet industry have acted as a double-edged sword. While the Internet industry has become a core business sector combining

Table 2.1 Global Entertainment and Media Market by Segment (millions of dollars)

	2004	2005	2006	2007	2008	2009	2010
Internet access	110,370	136,588	152,394	190,426	214,601	226,321	238,450
TV subscription and license fees	134,396	146,286	158,958	172,843	186,065	191,753	98,861
TV advertising	145,575	150,555	160,273	166,268	168,342	146,076	149,507
Recorded music	37,328	36,173	34,943	32,804	29,583	27,414	28,210
Films	82,834	80,833	82,233	83,869	83,925	84,833	91,045
Video games	27,807	29,815	34,504	43,460	51,390	56,089	58,383
Newspaper publishing	182,142	187,430	190,833	191,468	182,428	163,798	157,484
Magazine publishing	75,817	78,533	78,872	81,732	80,316	72,796	71,670
Radio	67,721	71,232	75,249	78,859	77,534	72,326	70,894
Total	863,990	917,445	968,259	1,041,729	1,074,184	1,041,406	1,064,504

Source: PricewaterhouseCoopers. (2009). *Global Entertainment and Media Outlook 2009–2013*, p. 33.

several old media functions, which has resulted in media convergence both technologically and corporately, the Internet industry has also been a major cause for the substantial decline in revenues of several old media industries, including broadcasting (both radio and television), newspaper, magazine, and music. On one hand, therefore, the Internet has become the future in the communication industries; one the other hand, the Internet has become a necessary evil for many media corporations, as will be detailed in later chapters.

CONVERGENCE IN THE GLOBAL COMMUNICATION INDUSTRIES

M&As in the media and telecommunications industries have rapidly increased in the midst of neoliberal globalization. As neoliberal globalization evolves, the transnationalization of the communication industries occurs daily, which means that many communication corporations, particularly in Western countries, have rapidly invested in non-Western countries as a form of M&As and joint ventures. These media TNCs have targeted emerging markets, such as China, India, Brazil, and Korea, in order to benefit from their emerging economies. Convergence in order to become a bigger media corporation is certainly beneficial in several ways. A bigger media corporation translates into product integration and control toward great reach, increased power, and ultimately a greater share of profits and larger profits from the world information and entertainment markets. As a result of convergence, content producers and distributors especially are coming under the same corporate roofs, even as these newly diversified behemoths concurrently transnationalized (Schiller, 2001).

On the one hand, vertical integration allows companies to keep the costs of sales and transactions low in that the vertically integrated corporations sell productions to themselves and, thus, do not have to go through bidding procedures and can have smaller sales and accounting departments (Gomery, 1986). This enables them to absorb the costs involved in introducing new products, which smaller firms cannot do, and therefore to compete more effectively in the emerging global market. On the other hand, vertical integration permits market control in that a corporation, which owns its own production companies and movie theaters or television networks, is guaranteed simultaneously a source of supply for its outlets and the exhibition or broadcast of the output it produces. Vertical integration in the communication industries therefore leads to the consolidation of corporate control over the market from the moment of a product's conception to its consumption (Gomery, 1986). In fact, vertical integration involves, for example, the ownership of film and television studios and/or production companies, distribution outlets, and movie theater chains and/or television networks. The trend of media consolidations highlights an increasing phenomenon of vertically integrated units within large media firms that aim to maximize profits by utilizing the resources and products from their own upstream or downstream holdings (Oba & Chan-Olmsted, 2006, pp. 99–100).

In contrast to vertical integration, horizontal integration secures corporate oligopoly in the communication sector. Horizontal integration involves the expansion of media firms in the same industry. For instance, broadcasting companies are acquiring more broadcasting firms to control the broadcasting market. As firms obtain a greater share of the market, they reduce their overhead and gain more bargaining power with suppliers. They also acquire more control over the prices they can charge for their products (*Financial Times*, 1998). As Robert McChesney points out (1999), firms operating in oligopolies—meaning markets dominated by a handful of firms each with significant market share—tend to do what monopolists do: They cut back on output so they can charge higher prices and earn greater profits. Many government legislations have prevented an extreme result of horizontal integration, which is the creation of a monopoly. There have also been restrictions on vertical integration insofar as maintaining competition. But for those companies that haven't yet established themselves among the mega vertically integrated players, the options appear to be running out.

Many broadcasting corporations, for example, have come under increasing pressure to expand their positions in order to remain globally competitive (Allbusiness.com, 2001). Vertical and horizontal integration secures these corporations' access to new markets and new sectors of the communication industries, the consequence of which is the oligopolistic control of much of the world communication market by a few giant corporations. Through vertical and horizontal integration, which is one form of M&As, media corporations have expanded their influence, and media ownership has been concentrated among several mega media giants.

CORPORATE CONVERGENCE AS A FORM OF M&A IN THE COMMUNICATION INDUSTRIES

According to the SDC Platinum database, there were 80,657 overall compiled M&As valued at $9,257.8 billion in the global communication industries, including the Internet, movie, broadcasting, advertising, and newspaper industries, during the period 1982–2009. Some of them overlap because they are connected through a cross-ownership structure, meaning newspaper companies have broadcasting companies. Therefore, in some cases, it is inevitable to see duplication in tallies. However, the important point is that these figures are good indicators showing the rapid growth of global M&As in the communication market, as well as domestic markets in many countries.

Among these, the Internet service and software industry (henceforth Internet industry) is the largest in both the number of deals and transaction values. Overall, 48,460 deals, valued at $4,634 billion, occurred as a reflection of the significance of the Internet in the deal market. As will be detailed in Chapter 7, most media and telecommunications corporations are eager to have the Internet sector as part of their segments, and this has resulted in the rapid growth of transactions in the Internet industry. Internet companies have also rapidly changed their ownership structure through convergence because many small Internet corporations cannot compete with Internet giants (Figure 2.1). Of course, the Internet industry has become a victim of the dot-com bubble since 2000 and has experienced a decline in transactions in the early 21st century.

The movie industry has been the second largest, followed by the broadcasting industry. These two audio-visual industries are under the same umbrella in several Western countries; however, they are separate in many non-Western countries, including Korea and China. This means that media corporations in the U.S. control the audio-visual sector, as in the case of Walt Disney and News Corporation; however, in many Asian countries, broadcasting and movie corporations are not connected, with a few exceptions, because they have been considered to be different fields. For example, in Korea, from the broadcasting firm's perspective, the film industry was not lucrative or attractive because it did not make profits for decades. Korean broadcasters therefore, unlike those

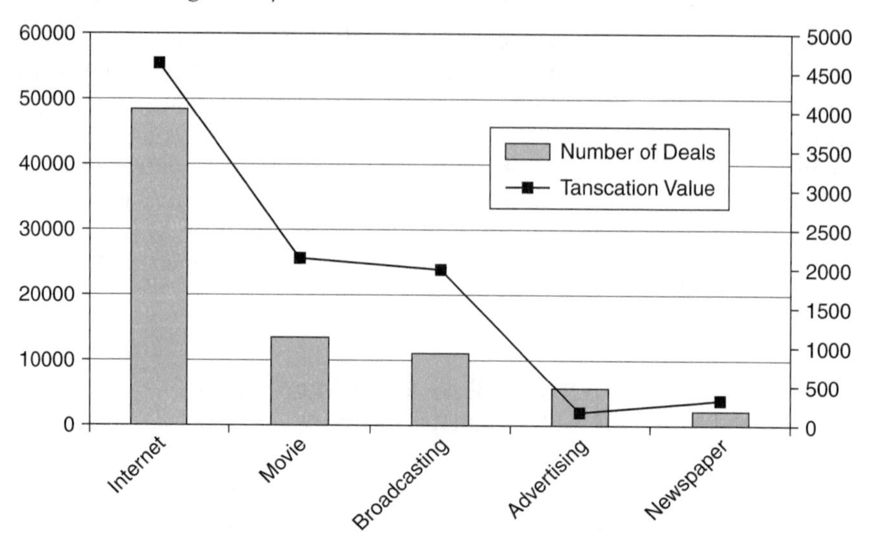

Figure 2.1 Convergence in the Communication Industries, 1982–2009 (billions of dollars)

in the U.S., did not have movie production companies (Jin, 2011b). Therefore, it is reasonable to separate them into two different industries. There have been 13,415 M&As, valued at $2,136 billion, in the movie industry, including production, distribution, and exhibition sectors, between 1982 and 2009. During the same period, the broadcasting industry, excluding cable corporations, recorded 11,055 deals, valued at $1,997 billion.

Advertising and newspaper industries are far behind in both the number of deals and transaction values. The number of deals in the advertising industry is 5,629, valued at only $170 billion, while there were 2,087 deals ($320.8 billion) in the newspaper industry. Given the size of the advertising industry compared to broadcasting or movie industries the number of deals in the industry itself is not small. The newspaper industry has been the smallest primarily because it has experienced severe declines in readership and in advertising revenues.

TRENDS IN MEDIA CONVERGENCE

The trend of convergence in the communication industries is very unique, and there are several major characteristics. Most of all, the amount of transactions was relatively low until the mid-1990s in most communication industries; however, deals in the communication industries have rapidly increased since then. As Table 2.2 shows, in the Internet industry, the number of transactions per year was not significant until 1995; however, it soared from 519 deals in 1995 to 8,037 in 2000. In the case of the movie industry, the number of deals was less than 500 per year until 1996; however, it soared to 836 in 2000. The broadcasting industry has shown a similar trend. The number of transactions in the broadcasting industry was 379 in 1994; however, it went up to 766 in 2000 (Table 2.2).

Media convergence as a form of M&A had mainly occurred between the late 1990s and the early 21st century. M&As completed in the com-

Table 2.2 Trends in Media Convergence, 1982–2009

Year	Internet	Movie	Broadcasting	Advertising	Newspaper
1982	23	41	50	8	13
1983	31	67	119	15	31
1984	37	100	140	36	38
1985	32	126	118	15	50
1986	76	266	189	51	65
1987	62	243	169	51	39
1988	84	371	203	160	53
1989	142	422	286	199	69
1990	162	360	200	159	53
1991	191	370	244	191	41
1992	181	341	239	163	59
1993	246	366	319	135	95
1994	362	401	379	182	79
1995	519	438	468	227	107
1996	759	537	490	241	108
1997	962	525	480	366	109
1998	1,735	584	496	432	89
1999	3,994	666	648	474	110
2000	8,037	836	766	475	130
2001	4,732	614	575	381	84
2002	2,973	559	457	250	55
2003	2,548	527	434	167	49
2004	2,812	601	481	240	73
2005	3,033	680	546	200	101
2006	3,601	834	630	223	110
2007	4,024	773	695	259	102
2008	3,970	889	669	189	95
2009	3,132	878	565	140	80
Total	48,460	13,415	11,055	5,629	2,087

munication industries between 1996 and 2009 accounted for as much as 85.6% of all M&As since 1982. In contrast to this, M&As completed until 1995 accounted for only about 14.4%. In particular, media convergence heavily occurred between 1996 and 2001. During only six years, 30,435 deals (37.7%) occurred. This means that slightly more than one out of three transactions during this time frame happened between 1996 and 2001. In fact, M&As in most communication sectors peaked in 2000; however, the deals have significantly decreased since 2001.

The data demonstrates that convergence in the communication industries has been fuelled by a few major historical events. Most of all, it is certain that a technological breakthrough in the late 1990s became a major factor for the growth of media convergence, particularly in the Internet and broadcasting industries, as well as other communication industries. After the commercialization of the Internet, many corporations and venture capitalists established their own Internet startups, and online businesses changed hands daily in the late 1990s. The old media industries, such as broadcasting and newspaper corporations, were especially interested in the Internet industry, and they vehemently pursued convergence strategies to acquire the Internet firms.

Meanwhile, newly expanding neoliberal economic and communication policies have become major factors for the growth of M&As. Privatization of communication companies has been noticeable during the last two decades. Typically, but not always, privatization is coupled with deregulation. With the collapse of the Soviet Union in 1991, the scale of the neoliberal privatization project in communication rapidly expanded and gained devotees within scores of countries.[3] Privatization within communication industries resulted in massive investing in the global communication market. Incumbents facing competition at home in mainly Western countries strongly pursued foreign markets for investment opportunities as a form of a consortium or joint venture (Frieden, 2001, p. 357).[4] In particular, one specific statute (the Telecommunications Act of 1996 in the U.S.) and one multilateral agreement (the WTO agreement in 1997), expedited the neoliberal transformation of the communication sector around the world (Schiller, 2003, p. 68). On the global level, the driving engine for neoliberal reform has culminated since the 1997 WTO agreement. The 1997 WTO agreement ignited a wave of M&As because each of the mega communication corporations moved aggressively into deregulated domestic communication markets around the world, and TNCs formed joint ventures with other global giants and local investors.

In fact, the U.S. enacted the 1996 Telecommunications Act, which allowed cross-ownership between telecommunications and broadcasting industries. The WTO agreement signed in 1997 also liberalized communication markets in many countries—in particular, non-Western countries. Therefore, both media and telecommunications industries have aggressively invested in the global communication markets, resulting in the surge of cross-border deals between two countries. Until the mid-1990s, cross-border deals were very limited; however, the situation suddenly changed in favor of transnational corporations who planned to invest in non-Western countries. Of course, in the midst of the liberalization process, a few emerging markets, such as China, Brazil, and Korea, have acquired communication corporations located in Western countries, although they are still very much marginal.

However, the number of M&As in the communication industry significantly decreased for a while in the U.S. and in other countries in the midst of shrinking global M&As right after the September 11, 2001 terrorist attacks on the U.S. As will be explained in detail in Chapter 5, the movie industry has been relatively less influenced by the economic recession, but other media and telecommunications industries have been severely hit by this historical event. In particular, the September 11 terrorist attacks, which became a major cause of the following economic recession in 2002 and 2003, has shifted the ownership structure of many communication corporations. Although several mega communication firms still pursue convergence, many of them do not pursue corporate convergence anymore. Instead, many communication firms, including several major media TNCs, such as AOL-Time Warner, Viacom-CBS, and AT&T, have sought a new business paradigm and newly adopted de-convergence strategies.

NEOLIBERAL TRANSFORMATION OF THE GLOBAL COMMUNICATION SYSTEM

The global trend toward transnationalization of the communication sector through M&As began with the rise of neoliberal globalization in the early 1980s, and, again, it has further accelerated with the 1996 Telecommunications Act and the 1997 WTO agreements since the late 1990s. A politics of neoliberal communication reform has driven dozens of nations to restructure their communication industries as a major part of economic transformation (Jin, 2008; 2011). The changes that were made in communication market structures in the 1980s indicated that competition, once the exception in communication, was quickly becoming the norm. The regulatory frameworks for the introduction of competition in telecommunications services differed. In the U.S., regional companies were set up, and a distinction had been created between local and long distance telephony; in Japan, a differentiation was made between facility owners/operators and other service providers, but competition was allowed throughout the network; in the U.K., although competition was allowed throughout the network, a fixed-link duopoly was guaranteed until at least 1990 (OECD, 1990, p. 7). As a result of a series of deregulating markets, transnational capitals were active in domestic telecom markets around the world. Transnational companies were intent on achieving access to an increasingly sophisticated, seamless communication network, enabling them to conduct business around the clock and around the world (Oh, 1996, pp. 714–715).

Capital investment occurred not only in domestic markets in developed countries but also in developing countries because communication giants extended their investment for high profits, reflecting their advanced technologies and saturation of domestic markets in developed countries, which resulted in the rapid growth of transnationalization of the communication sector. The strong trend toward the deregulation, privatization, and commercialization of communication, which began in the U.S. and the U.K., has opened up global commercial broadcasting and private telecommunications in a manner that displays a startling break with past practice throughout much of the world (Herman & McChesney, 1997, p. 45).

In fact, during the period 1982–2009, M&As in the same country (meaning the target company and acquirer are in the same country) accounted for 57,990 (72%), including unknown deals (about 5%), while cross-border deals comprised 22,656 (28%). Cross-border deals were not notable until the mid-1990s, but they have increased gradually since then. Between 1982 and 1995, for example, cross-border deals accounted for only 3,365 cases, consisting of 15% of the deals. However, the situation has changed. In particular, deals between countries have soared since 1997 with the WTO agreements. Cross-border deals between 1996 and 2009 were recorded at 19,291 (85%) out of an overall 22,656 cases. In 2000, cross-border deals peaked at 5,876 cases (26%), and there were 1,956 cases in 2003 (Figure 2.2).

The number of cross-border deals in the communication industries has greatly increased, and their characteristics are not much different from the overall trend. As a reflection of the boom of the Internet and related industries, the Internet and game software industries have been most active in the global deal market. Movie and broadcasting industries are beneath the Internet related industries in the category of cross-border deals. In fact, in

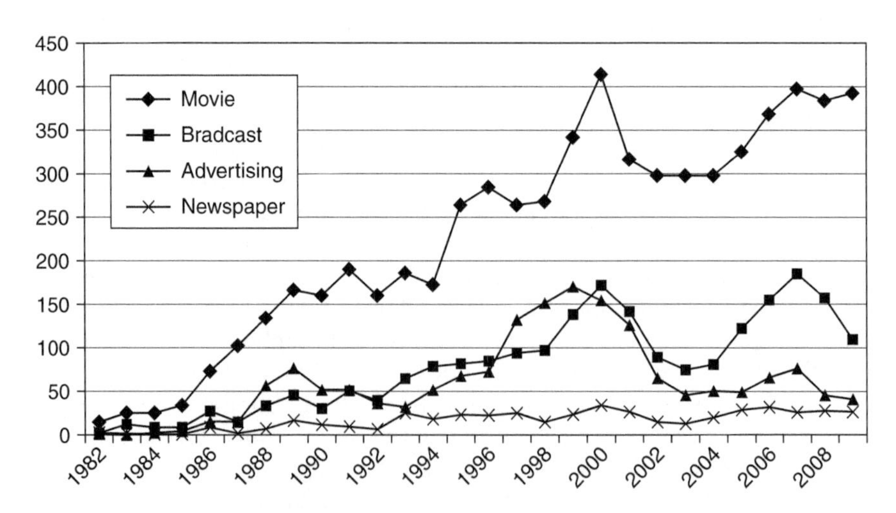

Figure 2.2 Cross-border Deals in the Communication Industry

2000, when cross-border deals peaked, the number of cross-border deals in the Internet industry was 2,162, followed by movie industry (415), broadcasting industry (171), advertising (156), and newspaper industry (34) (Figure 2.2). This trend has not changed much since the early 1980s.

However, in terms of the proportion of cross-border deals in each industry, the story is not the same. Among communication industries, the movie industry has been the most active in the global deal market. Overall, 6,348 M&As represent cross-border deals, which account for as much as 47.3%. This implies that almost one of two transactions in the movie industry occurred between countries, instead of within domestic markets. The advertising industry is the second most active at 30.7%, followed by the Internet industry (24.5%), newspaper industry (23.1%), and the broadcasting industry (19.8%).

This result implies several significant issues. Most notably, it is evident that the advertising industry has witnessed a rapid decline in cross-border deals so that it gives away the first position to the movie industry. The advertising industry used to be the most active communication sector in cross-border deals. During the period between 1983 and 2005, for example, the cross-border deals in the advertising industry accounted for 31.2%, followed by the telecommunications industry (24.7%), the newspaper industry (18.9%), and the broadcast industry (14.3%). The major function of the advertising industry has changed since the mid-1980s, from advertising only to dual functions (advertising and marketing), and mainly a few Western countries whose marketing skills were advanced had acquired advertising agencies from the developing countries in the late 1990s (Jin, 2008).

However, the number of cross-border deals in very recent years was not promising. In 1999, the number of cross-border deals in the advertising industry was 172; it dropped to only 42 cases in 2009. While other communication industries experienced a temporary decline in cross-border deals in the early 21st century, they have mostly recovered and maintained a similar number of deals in recent years. In fact, the number of cross-border deals in the movie industry peaked at 415 cases in 2000, and went down to 297 deals in 2004. It again went up to 392 in 2009. The situation is similar in the broadcasting and newspaper industries. Therefore, it is safe to say that the advertising industry has been experiencing a deep valley in the 21st century.

This data also confirms that the broadcasting industry has been mainly for domestic audiences; therefore, broadcasting corporations in many countries target domestic markets. Although cross-border deals in the broadcasting industry have been stable with a few exceptional years (for example, 138 cases in 1999, 122 cases in 2005, and 158 cases in 2008), these are not comparable with the transactions that occurred in domestic markets. It is expected that the broadcasting industry had the lowest cross-border deals in the communication industry, because television remains primarily a national phenomenon in many countries, and the domestic state plays a significant role in shaping national television systems (Straubhaar, 2001, pp. 133–134).

Considering the social and cultural impact of broadcasting services on the public, foreign ownership restrictions on broadcasting services were common in many countries in the 1980s (OECD, 1999, p. 114). Many countries, from the U.S. to Asian and Latin American countries, however, have selectively (partially, if not wholly) begun to lift bans against foreign ownership in national communication industries beginning in the mid-1990s, although traditions of protection for domestic communication and cultural industries persist. In other words, M&As in broadcasting stations mainly occur between domestic companies because broadcasting companies are symbols of national identity, but with the relaxation of foreign ownership restraints and privatization, cross-border M&As have gradually increased in recent years (Jin, 2008).

The general assumption regarding the newspaper industry is that newspapers are a text-based medium, so newspaper companies are interested only in domestic markets. However, the current results show that the newspaper is not inactive anymore. Cross-border deals in the newspaper industry account for 23.1%, which is very close to the Internet industry (24.5%). That means that almost one fourth of deals in the newspaper industry occur between two different countries. There are several big English-based newspapers, as in the case of News Corporation, and they have expanded their influence into other English-speaking countries, which is the major reason for the cross-border deals in the newspaper industry.

Meanwhile, the number of cross-border deals in the telecommunications industry has substantially increased from only 5% in 1983 and 4% in 1984 to 32% in 1999. The amount of cross-border deals in the telecommunications sector and the broadcasting sector have further increased since 1996 and 1997, mainly because of the 1996 Telecommunications Act and the 1997 WTO agreements. The most profound impact on the global telecommunications markets came from the concluded multilateral negotiations on basic telecommunications services in WTO under auspices of the General Agreement on Trade in Services (GATS). The basic telecommunications negotiations were launched formally in 1995 after the establishment of the WTO. The launching of basic telecom negotiations at the WTO sent a strong message to the telecom officials and telecom companies across the country. The Agreement on Basic Telecommunications Services requires them to open their domestic markets to foreign competition and to allow foreign companies to buy stakes in domestic operators (McLarty, 1998, p. 4; Jonquireres, 1997, p. 6). As a result of these two major historical events, the amount of cross-border deals in the telecommunications industry increased from 23% in 1995 to 32% in 1999 and to 31% in 2000. In the broadcasting industry, the ratio increased from 14% in 1995 to 19% in 2000.

The two historical events have also changed the landscape of the global communication system because the U.S. case has demonstrated a contradictory phenomenon in the deal market in recent years. As discussed, the role of the U.S. communication industries in the overall deal market has greatly increased over the past two decades, and the country has intensified its power, particularly since the late 1990s, in the global communication system. The neoliberal reform movements, first in the early 1980s and second in the late 1990s, which were driven by the U.S., have changed the map of the global communication system via massive capital investment in a form of M&As. U.S. communication companies have enjoyed supreme power in the midst of the neoliberal reform, not only domestically, but also in other countries, by investing their capital.

CONCLUSION

The global communication system has substantially changed since the mid-1980s. The communication industries have grown dramatically with worldwide economic restructuring and technological innovations. The changing political-economic environment in the world caused the rapid transformation of global communication. The introduction of neoliberal economic policy, adopting deregulation, liberalization, and privatization of communication systems beginning in the mid-1980s in the U.S. and the U.K., has resulted in the breakup of public monopolies over broadcasting and telecommunications industries, as well as other communication industries. Privatization and the relaxation of foreign ownership restraints have expedited the swift transnationalization of the global communication system.

The restructuring of the communication industries since the mid-1980s was first made possible because governments around the world adopted deregulation and privatization polices, as they confronted intensifying pressure not only from corporations but also directly from international organizations and the U.S. government. In particular, the WTO agreement on basic services attracted widespread attention with many governments because it succeeded, on a global scale, in establishing the free trade principle in an area previously closed to foreign intervention. In 2000, China, for example, declared that it would partially open its communication market toward international communication as part of China's WTO accession. With the lifting of the ban on the global communication industry, several foreign majors invested in the Chinese communication market, including News Corporation and Disney. As Yuezhi Zhao points out (1998, pp. 170–172), external challenges to the Chinese media system have increased since the 1990s.

Not surprisingly, TNCs played a significant role in the process of change, along with international organizations and the U.S. government. With deregulation in each government, TNCs have invested an enormous amount of money in the communication industry in developed countries because it became a highly profitable sector of the world economy, which resulted in developing at ever increasing rates of growth in the communication industry. In other words, commercialized mega communication and telecommunications industries through M&As can control the global market, and ultimately acquire a larger share of and larger profit from the global information and entertainment markets. The emergence of mega communication companies was driven to allow large companies to control content and hardware together in order to enable them to maximize their value and profit. Mega communication conglomerates can secure the outlet of their content, including television programs and films, through vertical and horizontal integration, as well as through international alliances.

However, the role of developing countries has rarely increased in the global M&A market over the last decade. It is argued that new forces, particularly several local-based transnational communication companies have expanded their role in the global market in the process of the global integration. Keane (2006) pointed out that neoliberal globalization does not solely mean the dominance of Western countries. The increased level of transnational information flows, made possible by the new technologies of communication and shifts in the institutional organization—economic, political, and legal—on the means of communication, have profoundly affected global media industries. Wang (2003, pp. 267–268) states, "although a few developed countries have increased their roles in the global communication market, with scarce resources and few competitive advantages, many developing countries fought bitter battles to defend their sovereign rights when issues regarding transborder data flows and direct satellite television broadcasting emerged on the agenda of international negotiations."

With integration of content origination through to delivery mechanisms, non-U.S. based TNCs have functioned in the global communication industries; however, the key players in the era of neoliberal globalization are still mega corporations in Western countries. The role of the U.S. has slightly decreased; however, only a few developed countries have accounted for the majority of M&As in the communication industry. The growth of the developing countries in the global deal market is far away. Although some developing countries, including Mexico, Singapore, and Korea, have increased their roles in the cross-border deal market, the inequality and imbalance in the communication industry between developed countries and developing countries still exists in the midst of neoliberal transformation. Although the U.S. has slightly lost its power in the global M&A market, the gap between a few Western countries and developing countries remains quite a distance.

3 Transformation of the Global Broadcasting Industries

Television has been one of the most significant entertainment tools and the most reliable news distributor since its inception. As our society has focused on more visual-driven mass media, television has functioned as the core of the media industry and audio-visual culture for everyone. From the early black and white television era to the very nascent Internet television era, television is a major part of everyday life. It is certain that several new media and social media, including the Internet and social networking sites (SNSs, e.g., Facebook), have rapidly grown and replaced some TV functions. As the survey of 50 countries found, people who watched television indeed declined 7% in 2011, primarily because people who watch video content on computers increased (Nielsen, 2012); however, television viewing still remains popular, and television will continue to influence our culture and economy in somewhat different degrees in the 21st century.

While television viewing has been a significant part of our daily activities, the television industry has undergone a dramatic change as the primary component of global communication industries over the last several decades. With the rapid growth of the economy, the number of TV sets and channels has soared. The swift development of new technologies has also changed traditional distribution of audio-visual content, from mainly terrestrial channels to multi channels, including cable, satellite, and digital channels. For both network and cable television companies, finding a workable strategy in an increasingly digital environment has been a significant issue, because they worry that television's role in the media landscape could diminish as Internet TV and social media, including YouTube, are taking a more important role as aggregators and distributors of information and entertainment content (Dennis, 2006, p. 22). The broadcasting industry therefore has been privatized and commercialized, and it has developed its convergence and later de-convergence strategies in order to survive amid a changing media environment.

Since the mid-1980s, the broadcasting industries in a few Western countries have indeed vehemently expanded their influences to both Western and non-Western nations with their massive capital, technologies, and television programs. They have especially invested in domestic broadcasting industries in non-Western countries as major stockholders through M&As and/or in the form of joint ventures. These major transnational media corporations have become primary forces behind the swift growth of non-Western broadcasting industries around the world. Along with the growing roles of the transnational corporations, several other factors also impinged on the rapidly changing global broadcasting system, including continuing worldwide economic transformation and technological development (Jin, 2007).

This chapter historicizes some of the key features of the global TV system that has been reorganized since the mid-1980s. It maps out the changing structure of the broadcasting industry by examining convergence driven by neoliberal reforms, including privatization and liberalization. It also discusses whether the nation-state has taken a pivotal role in the transformation of the broadcasting industry amid neoliberal globalization.

A SHIFTING GLOBAL BROADCASTING SYSTEM AMID NEOLIBERAL GLOBALIZATION

The broadcasting industry has been a core business in the communication industries. Due to its increasing influence as an entertainment and information medium and cultural icon, broadcasting has become the most wanted media business in the communication industries. The emergence of the global TV industry as a strategic and highly profitable sector of the world economy and the rapid development of new transmission and distribution technologies—including satellite, cable, and VCR—especially spurred a huge increase in television outlets, fuelling explosive trade growth in television programs and films (Jin, 2007). The rapid transformation of the global broadcasting system has been possible because governments around the world adopted neoliberal communication policies, such as market deregulation and reduced state intervention in communication affairs in general, as they confronted intensifying pressure from TNCs and the U.S. government (McChesney, 2008; Schiller, 1999a, 1999b), and the neoliberal communication policies have become an indispensable element for consolidation and reform within the broadcasting industry.

There are several dimensions reflecting the swift change of the global broadcasting system, including the number of television sets and channels, as well as the broadcasting ownership patterns. To begin with, due to its significance to people as a cultural icon, mega media companies around the world have focused on television as the/a major media business, although some of them are vertically and horizontally converged with other media industries, including Internet businesses. All of the top 15 global media companies have television; however, five of them are solely operating a television business. Among these, News Corporation, the largest media company in the world, made $30.4 billion in revenue in 2009, and it has newspaper, magazines, books, radio, and film businesses, as well as Fox TV.

Table 3.1 Top 15 Media Companies in the World in 2009

	Company	Country	Revenue (U.S. $M)	TV	Newspaper	Magazine	Book	Radio	Film	Music
1	News Corporation	U.S.	30,423	x	x	x	x	x	x	
2	Time Warner	U.S.	25,785	x		x	x		x	
3	Walt Disney	U.S.	23,057	x		x	x	x	x	x
4	Comcast	U.S.	22,317	x						
5	DirecTV Group	U.S.	21,565	x						
6	Bertelsmann AG	Germany	21,343	x	x	x	x	x	x	x
7	NBC Universal	U.S.	15,436	x				x	x	
8	Viacom	U.S.	13,619	x					x	x
9	Vivendi	France	12,354	x						x
10	Dish Network	U.S.	11,663	x						
11	Time Warner Cable	U.S.	11,462	x						
12	CBS Corp.	U.S.	11,292	x				x	x	
13	Lagardere	France	10,963	x	x	x	x	x	x	
14	Cox Enterprises	U.S.	10,800	x	x			x		
15	Liberty Global	U.S.	8,953	x						

Source: NORDICOM (2010). *A Sampler of International Media and Communication Statistics, 2010*. Gothenburg, Sweden: Nordicom.

Time Warner and Walt Disney have also operated several media businesses, including television. However, among top 15 global media corporations, Comcast (4th), DirecTV Group (5th), Dish Network (10th), Time Warner Cable (11th), and Liberty Global (15th) own only television. Of course, these top 15 mega corporations are mostly located in the U.S., with three non-U.S. corporations: Bertelsmann AG (Germany), Vivendi (France), and Lagardere (France) (NORDICOM, 2010). (See Table 3.1.)

For media corporations, television has been the most important medium due in large part to its appeal to people as an audio-visual medium, which is one of its most valuable assets. With the rapid growth of new broadcasting technologies and channels, such as cable and satellite, several new media corporations have focused on these new media and have themselves become media giants.

Meanwhile, the broadcasting system has witnessed rapid growth in several areas. The number of television sets, including television sets in use at home, in businesses and in hotels in the world has exponentially soared, from only 299 million in 1970 to 1.62 billion in 2000 (ITU, 2002; UNESCO, 1999). Despite competition from the Internet and mobile technologies, the number of television sets has been growing mainly because of the explosion of television sets in many developing countries in Asia and Latin America. The growth in the television market in these countries has been distinctive since the early 1990s, due to economic growth and the introduction of new technologies, such as satellite and cable television. In addition, the television market has shown saturation in several developed countries, and it is not growing anymore in these countries.

By 2009, there were some 1.4 billion households around the world with a TV (this number excludes businesses). This resulted in a household penetration of 79%, up from 73% in 2002. Europe and the Americas had a household television penetration of over 90%, while in the Arab States and Asia and the Pacific, penetration stood at 82% and 75%, respectively. Africa, where 28% of households had a TV, stands out for having the lowest levels of household TV penetration (ITU, 2010, p. 155). By country comparison, as of December 2009, China, based on its largest population and its emerging economy, had the largest number of households with TVs (385.7 million), followed by India (140.1 million), the U.S. (114 million), Russia (50 million), and Japan (48.3 million). While these countries also had higher penetration rates of over 95%, many countries stood at less than 70%, such as India (64%) and Indonesia (60%), and low penetration was evident in Africa and Latin America (NORDICOM, 2010).

Since the late 1990s, television and video content has been available on a number of different types of networks and can be viewed on many different devices, giving viewers an expanding choice of personalized viewing options. The increasing availability of audio-visual content on alternative networks, whether terrestrial television, cable, Internet Protocol Television (IP TV), or mobile phones, is particularly notable (OECD, 2011, p. 220). In other words, a noteworthy trend over the last decade has been the development of multichannel television. The number of households with multichannel TV[1] had risen from 400 million in 2000 to 650 million in 2008. Around two out of five television households were multichannel in 2000, compared to almost half by 2008. The increase in multichannel TV has contributed to an increase in content, which in turn has increased the demand for television services, as exemplified by a number of countries in South Asia (ITU, 2010). Cable television (CATV) is the leading multichannel option, with 448 million CATV households around the world by 2008. China is by far the largest CATV market, accounting for some 64% of the world's subscribers. Cable is well developed in the Americas, Asia and the Pacific, and European regions, but virtually nonexistent in the Arab States and Africa (ITU, 2010). About 43% of Chinese households had multichannel TV in 2009, followed by Japan (24%), Vietnam (17%), and Indonesia (4%). Multichannel TVs are relatively less common in Asia (NORDICOM, 2010). In the Asian region, Korea (99%) and Taiwan (99%) show the highest penetration rate of multichannel TV. While the multichannel system has rapidly become popular in North America and Europe, many Asian and Latin American countries are still struggling to catch up with these Western counties, implying another form of digital divide, this time between those who have multichannel televisions and those who do not have them.

The surge of multichannel television has changed TV viewing habits in several Western countries. Western countries already experienced the saturation of television sets in the early 1990s, reflecting their full-grown television service markets. While the increase in television sets in China has been the steepest since the 1990s, the ownership of television sets in the U.S. fell in 2010, for the first time in 20 years. As Nielsen reported, as of December 2010, 96.7% of American households own TV sets, down from 98.9% from the previous year. Two reasons are cited as major reasons for the decline. One is poverty. Some low-income households no longer own TV sets, most likely because they cannot afford new digital sets and antennas. The other is technological wizardry: Young people who have grown up with laptops in their hands instead of remote controls are opting not to buy TV sets when they graduate from college or enter the workforce, at least at first. Instead, they are subsisting on a diet of television shows and movies from the Internet (Stelter, 2011). Since this phenomenon has become real in many emerging countries, the number of TV sets sold might continue to decline throughout developing and a few emerging markets.

Therefore, it is not surprising that the number of television channels has proportionally grown over the past several decades. For example, the total number of television channels available in European countries soared to 7,918 as of December 2008 (NORDICOM, 2010). The total number of television channels has been increasing rapidly due to satellite and cable television networks. Satellite and cable broadcasting companies have been at the source of the creation of many new television networks around the world, and they have served as catalysts in the globalization of television, although terrestrial television systems also increased gradually (Jin, 2007).

The growth of multichannel television throughout the world and, in particular, in several Asian and Western countries, has resulted in the growth of foreign television programs, not only from regional broadcasters but also from U.S.-based firms. Many new channels, including on network, cable, and satellite television, could not produce their own programs due to lack of production money and technologies. As a result, they have to seek television programs elsewhere, and the U.S. broadcasting corporations, alongside a few local providers, including Brazilian and Korean broadcasters, have certainly benefited from the growth of the global television market. In other words, the rapid growth of the regional television market has facilitated the flow of television programs and television sets between the West and the East; therefore, the process has become a major part of the globalization process.

STRUCTURAL CHANGES IN THE GLOBAL BROADCASTING SYSTEM

The phenomenal growth and change in the global broadcasting industry encompasses several key characteristics: the emergence of mega global communication companies as a result of mergers and acquisitions; privatization of existing broadcasting companies; the relaxation of foreign ownership restrictions; corporate investment in newer media, such as cable and satellite television; and transnationalization of advertising and its convergence with communication empires to create a demand for and to promote cultural products and other industries, such as consumer goods and services (Jin, 2007).

To begin with, the rise of the global broadcasting industry resulted from corporate convergence to enhance profits in the global media market. For example, the French media and utilities company Vivendi bought Seagram, the owner of Universal Studios and Polygram Records, for $34 billion in 2000. The move marked the transformation of Vivendi into a global new media player, and the deal had been driven by the need for companies like Vivendi to acquire content, such as music albums and films, which they can sell over the Internet. The purpose was "to make the internet swing," according to Vivendi chief executive Jean-Marie Messier (BBC News, 2000). However, the convergence mostly failed, and Vivendi had to sell several of its media units, including Vivendi Universal Entertainment, to NBC, because of the group's huge debts (Noble, 2003). NBC purchased Vivendi Universal Entertainment because of the synergy effects from this integration. NBC expected to utilize its convergence for the cross-marketing of its television, movie, and theme park properties to drive profits (Carter, 2004).

NBC's bid to control diverse outlets through corporate convergence has unluckily failed as well. NBC has been mired in fourth place among the major broadcast networks in the U.S., and the economics of broadcast television has deteriorated in recent years amid declining overall ratings and a decline in advertising. NBC had no choice but to sell itself to another media giant, Comcast, the largest cable company. Cable channels have continued to thrive because they rely on a steady stream of subscriber fees. The combination of Comcast's cable systems and NBC Universal's channels created a media powerhouse, and it represented the first time that a cable company controlled a major broadcast network (Stelter, 2011).

In the U.S., the recent acquisition of the Fox Family cable channel, the Saban library, and Fox Kids International by Walt Disney is another textbook example of utilizing vertical integration. Domestically, Fox Family Channel gives Disney 81 million cable homes, but the real cornerstone of the deal was Fox Kids International. Disney has already branded broadcast blocks and 14 Disney Channel premium services around the world. Disney has 600 million people per month watching those assets, but with the basic advertiser-supported channels of Fox Kids, which reach 35 million people worldwide, Disney expected the synergy effect of vertical integration. Meanwhile, the horizontal integration of Fox Family and Fox Kids into the existing Disney cable channels added much needed cash flow (Allbusiness.com, 2001).

As such, through M&As, media empires such as Viacom, Time Warner, News Corporation, Walt Disney, and Sony have rapidly expanded their dominance in the global communication market through both vertical and horizontal integration, including their operations of the production of television programs and films, distribution, and exhibition. As discussed, television has been one of the most significant media for mega media giants. Through their terrestrial and/or cable channels, they produce content and exhibit their programs. In order to control these markets, they have vehemently pursed vertical and horizontal integration, because for them 'big is beautiful.'

CONVERGENCE OF THE GLOBAL TV INDUSTRY

The convergence paradigm in the broadcasting industry as in other media sectors has a long history. However, the trend toward transnationalization and conglomeration of the broadcasting industry through corporate integration mostly started in the 1980s when the U.S. broadcasting industry experienced a spate of corporate takeovers. The TV station industry saw its concentration grow significantly in the period between 1984 and recent years, enabled by the loosening of ownership caps set by the Federal Communications Commission (FCC) (Noam, 2009). In other words, in 1985, the FCC changed its media policy, enabling large corporations to buy smaller enterprises (Gomery, 1986). In 1985, the FCC also raised the number of television stations a single company could own from 7 to 12 (Tucker, 1985).

The U.S. government, through the FCC, cleared the way for an ambitious and wealthy person like Rupert Murdoch to acquire vast concentrations of power (Gomery, 1986).

Consequently, ABC was bought by Capital Cities Communication, and NBC was bought by General Electric (GE) (Stevenson, 1985). When ABC agreed to be acquired by Capital Cities Communications, it was the first time that one of the nation's three major television networks changed hands since Leonard H. Goldenson's United Paramount Theaters merged with a fledgling network to create the modern ABC in 1953 (Stevenson, 1985). The same year, CBS was beset by Ted Turner's unwelcome takeover bid (Stevenson, 1985). In April 1985, Ted Turner, the chairman of the Turner Broadcasting System, began a hostile takeover bid for CBS, which struggled with huge losses. CBS was eventually acquired by Loew's, a medium-size media company. Capital Cities Communication also acquired ABC in March 1985 in a transaction valued at more than $3.5 billion, which was the largest acquisition in U.S. broadcasting history until then (Storch, 1985). These near simultaneous changes in network ownership in 1985–1986 passed control of the network from the broadcasting industry's founder-owners to more conventional corporate management. As usual, when a privately held company is acquired, asset ownership is separated from managerial control, although the former owner-managers of the company that is sold may stay on with the acquirer as executives and will often have quality stakes in the acquirer as a result of the sale (Lazonick, 2009, p. 222). The new owners were more cost conscious than their predecessors. For example, in its first year of control, Capital Cities cut ABC's costs by $100 million by firing 1,600 people from ABC-owned TV networks (Noam, 2009). Rupert Murdoch, the founder and owner of News Corporation, also began to create another major television network, Fox TV, which started in October 1986, to be a key player in the global broadcasting market onwards. Once these major networks changed their major ownership structures, several of the nation's largest television stations and station groups also followed the trend, and they were major targets in the domestic M&A market. In fact, a politics of neoliberal media reform has taken hold in most countries since the mid-1980s. While there were no significant global M&As in the broadcasting industry until the early 1980s, convergence in the broadcasting industry swiftly increased in the 1990s. M&As among broadcasting stations between 1982 and 2009, in particular, showed a unique trend in the international M&A market.

According to the SDC Platinum database, 11,055 overall M&As, valued at $1,996.7 billion, were completed worldwide among broadcasting companies between 1982 and 2009 (Table 3.2). Among these, U.S. broadcasting stations acquired as many as 4,719 broadcasting stations (42.6%), including acquired companies that were either U.S.-owned or foreign owned, followed by the U.K. (784 companies), Canada (619), Australia (487), Germany (390), Japan (343), and France (324). These seven countries acquired

Table 3.2 Mergers and Acquisitions of the Broadcast Industry

Year	Numbers of Deals	Transaction Values (millions of dollars)
1982	50	1,291.5
1983	119	2,858.8
1984	140	7,544.8
1985	118	32,825.2
1986	189	46,968.6
1987	169	23,273.4
1988	203	28,766.5
1989	286	41,682.5
1990	200	18,734.0
1991	244	17,142.9
1992	239	9,697.1
1993	319	26,959.8
1994	379	94,098.3
1995	468	49,386.2
1996	490	70,945.5
1997	480	62,300.4
1998	496	85,576.8
1999	648	146,647.9
2000	766	195,936.2
2001	575	156,671.1
2002	457	112,629.6
2003	434	53,550.5
2004	481	132,404.5
2005	546	104,343.3
2006	630	152,690.1
2007	695	109,571.3
2008	669	146,930.2
2009	565	65,349.9
Total	11,055	1,996,776.9

7,666 companies (69.3%), and they dominated the global M&A market. The relatively high number of deals made by Australian companies was a result of the aggressive M&A strategy of News Corporation, owned by Rupert Murdoch.

The majority of M&As have occurred in recent years. M&As completed among broadcasting stations between 1996 and 2009 accounted for 71.9 %

(7,932 deals) of all M&As since 1982. M&As among broadcasting stations peaked in 2000 with 766 deals; however, the number of deals substantially decreased from 2001 in the midst of the decline of the global M&As—with 575 cases that year and 457 cases in 2002—and then bounced back with 695 cases in 2007 and 669 in 2008. M&As in the broadcast industry decreased significantly for a while in the U.S. and in other countries in the midst of shrinking global M&As in almost all industries, after the September 11, 2001, terrorist attacks on the U.S. The number of global M&As in the broadcasting industry increased for a while between 2006 and 2008; however, it could not match the peak point of 2000, due to the current economic recession mainly started in the U.S.

During the same period, M&As in the same country (meaning the target company and the acquirer are located in the same country) accounted for 8,496 cases (79.5%), while cross-border deals comprised 2,192 cases (20.5%).[2] Cross-border deals were not significant in the 1980s, but increased gradually in the 1990s. Majority mergers and acquisitions have also occurred in recent years. M&As completed among broadcasting stations between 1999 and 2009 accounted for 65% of all cross-border deals since 1982. This data demonstrates that M&A in broadcasting stations mainly occurs among domestic companies because broadcasting companies are symbols of national identity, but with the relaxation of foreign ownership restraints and privatization, cross-border M&As have gradually increased since the late 1990s.

By all accounts, local television in the U.S. is relatively robust once again. Automotive advertising is riding high, and retransmission consent is a consistently bankable revenue source. However, it is not without considerable risk due to an ongoing recession as of April 2010, and no one can seem to agree on what TV stations are worth anymore (Malone, 2010). The slow activity of M&As in the broadcasting industry has been recovered, but the gap between buyers and sellers still remains substantial.

TRANSFORMATION IN THE BROADCASTING INDUSTRY

The structural change in the broadcasting industry as a form of corporate convergence has been made possible with neoliberal media policy. Considering the social and cultural impact of television on the public, foreign ownership restrictions on broadcasting services were common in many countries in the 1980s (OECD, 1999). However, the situation has shifted since many developing countries partially, if not wholly, have been lifting bans against foreign ownership in their national broadcasting industries since the mid-1990s, although traditions of protection for domestic communication and cultural industries persist. As of 2007, there were no restrictions on foreign ownership, in both terrestrial and cable television, in most European countries, including the U.K., Belgium, Ireland, Denmark, Luxembourg, the

Netherlands, Norway, and Sweden and in a few other countries, such as New Zealand, and among OECD countries (OECD, 2007). Many countries allowed foreigners to own 20%–49% of shares in the broadcasting industry, as in Brazil, which used to prohibit foreigners from owning stakes for the development of domestic companies and their programs (Jin, 2007). Ironically, the U.S. and Canada still limited foreign ownership to 20% in all broadcasting sectors, including terrestrial, cable, and satellite systems (OECD, 2007).

More specifically, in 2004, the Korean government amended its broadcasting law and allowed foreigners to own up to 49% of shares in cable and 33% in satellite, although the government still regulates the terrestrial broadcasters from this liberalization measure (Korean Broadcasting Law, Article 14.3). The government wanted to ease the financial hardships of cable networks through foreign ownership. When the government planned this amendment in the late 1990s, the government announced that "this is the general trend. With the nation entering into a global competition era, we can no longer cling to the old practice of protecting certain sectors in the economy" (*Korea Times*, 1998). Foreign media firms certainly saw great advertising and consumer spending potential of this emerging market and acquired stakes in several cable companies (Jin, 2011a). Prior to this, the government also partially allowed the ownership of foreign corporations in digital satellite television. According to the Integrated Broadcasting Act passed in 1999, foreign interests could own up to 33% of shares in satellite broadcasting companies. As a result, foreign investors gradually penetrated the Korean broadcasting market. Capital International (U.S.) also invested $50 million in On Media, a Korean broadcaster owned by Dong Yang, and it became the second largest stockholder (21.68%) in 2000, while New Asia East Investment Fund Ltd. (Singapore) also invested in On Media (5.42%) (Cho, 2002, p. 114). On Media was absorbed by CJ Group in 2011.

The most crucial transformation came from China. In 2000, China declared it would partially open its communication market toward international communication as part of China's WTO accession, which was finalized in 2005. China, which had rejected foreign investment and foreign ownership in the communication industry until the late 1990s, permitted foreign companies to own up to 49% of video and audio distribution companies and up to 49% of companies that build, own, and run cinemas (Hazelton, 2000, p. 8). China's largely insulated economy continued to experience double-digit growth. China's acceptance of global market rules coincided with its entry into the World Trade Organization in December 2001 and was fundamentally a triumph of pragmatics over national sovereignty. Entry into the world's premier trading club suggests a force that smashes old institutional practices and allows the marketplace to rebuild with greater capacity (Keane, 2006). With the lifting of the ban on the global communication industries, several foreign media giants, including News Corporation and Disney, invested in the Chinese communication market; Viacom first entered

China through its entertainment arm MTV Asia, which covers three regional channels: MTV Mandarin, MTV Southeast Asia, and MTV India (Fung, 2006). China's global integration through the communication and cultural industries, including the scope and pattern of foreign media penetration, has become common partially in return for its entry to the WTO system (Zhao, 2008, p. 35). The role of transnational capital flows and communication networks undermined the power of the nation-state, although the Chinese government tried hard to contain the influence of the foreign media out of fear of ideological influence from the West (Zhao, 2008).

The privatization of the broadcasting industry has also been one of the main features of the global communication system over the last two decades. Privatization is the transfer of property and/or operations from state or public ownership and control into private hands. Among the principal reasons given to justify privatization is that private ownership and operation make a company perform more efficiently because its managers will be financially obligated to make the company accountable to shareholders (Carabrese, 2008). In the midst of neoliberal globalization emphasizing the minimal role of the government in economy and culture, the privatization of public broadcasting systems has become a common routine in many countries. Corporately run systems have sprung up in dozens of countries; elsewhere, existing private systems have been expanded and enlarged (Schiller, 2001).

Meanwhile, the development of new channels, including cable and satellite television channels, as well as terrestrial, in Europe, Latin America, and Asia has expedited the opening of national television program markets for foreign producers and distributors, in particular those of the U.S, but partially those of regional producers. Technology, along with the rapid growth of the regional economies, is an accelerating force in developing television programs and film trade in several regions, including East Asia (Jin, 2007). Several countries around the world also lifted a ban against foreign television programs in terrestrial television, and to a higher degree in cable and satellite television. Although several countries still restrict foreign programs in their terrestrial channels, many countries did not require any quotas for domestic or local programs in cable television. A few countries indeed try to block foreign television programs and films in both terrestrial and cable channels in order to protect their own broadcasting industries. For example, the Chinese government asks broadcasting channels not to air foreign programs during so-called prime time between 7:00 PM and 9:00 PM as of February 2012 (Simpson, 2012). The Taiwanese government also took a very similar protection measure. In December 2011, the National Communications Commission asked several cable channels to adjust their prime-time programming because the number of hours for Korean dramas was too much. The commission clearly asked them to increase non-Korean dramas from one hour to two hours between 6:00 PM and midnight on weekdays (Shan, 2011). These protective measures are very rare in most countries, although China and Taiwan seem to target burgeoning Korean dramas in their channels. In the midst of the rapid growth of Korean cultural products and their export to East Asian countries, Korean programs have become very popular in these countries; therefore, these governments unusually regulate their broadcasting markets. With these few exceptions, the global television market has opened for foreign programs.

In fact, several emerging forces have played key roles in the global broadcasting system, including Asia, Africa, and Latin America, in recent years. Emerging domestic broadcasting corporations, including CCTV in China, *Televisa* in Mexico, *Teleglobe* in Brazil, and MBC and KBS in Korea, as well as more national-based TNCs in these countries play significant roles in the process of changes in the global broadcasting industries. The role of emerging domestic players as both producers and distributors has been gradually increasing, as in the case of the Korean Wave, which is symbolizing the sudden growth of the Korean cultural industries, including television programs, although that does not mean that the imbalance and inequality in the broadcasting system between the West and the East is reduced fundamentally.

Of course, the beneficiaries of deregulation in the broadcasting program market are a few Western countries, mostly the U.S. According to the U.S. Department of Commerce, U.S. film and television program exports in current dollar terms as a form of royalties and license fees were valued at slightly over $1 billion in 1985 and $2 billion in 1990. However, the U.S. exported about $8.5 billion worth of film and television programs to the world in 2001, and this amount soared to as much as $13.5 billion in 2010 (U.S. Department of Commerce, 2011; 2002). The U.S. exports increased 13 times between 1985 and 2010. However, the U.S. imported only $123 million worth of television programs and films from other countries in 2001, and it was still only $1.67 billion in 2010 (U.S. Department of Commerce, 2011).[3] The net profits of the U.S. in this particular service category reached $11.86 billion in 2010. Until the early 21st century, most of this large jump could be attributed to increased exports of U.S. television programs to new cable channels in foreign countries, and later digital channels, because these countries have not accumulated skills and stories to create new programs for these burgeoning channels.

This data implies that there is no doubt that a few Western countries, including the U.S., have maintained cultural power in terms of the magnitude of the trade in television programs around the world. There still seems to be more of a potential for uneven flow of sales. Every world region, including Asia, imports far more cultural products from the U.S. than it sells to the U.S. Government deregulatory policy around the world has expedited the global trade of audio-visual product flow, along with other factors, such as technology and culture (e.g., ethnicity and language) (Jin, 2007). Although several developing countries have increased their global presence with their capitals and programs, a few Western countries, the U.S. in particular, have dominated the global broadcasting market with their advanced know-how, programs, and capitals. They have acquired broadcasting companies in non-Western countries, and they have provided television programs, both news and entertainment programs. The rapidly changing global broadcasting industries have been a result of the interaction between the West and the East—among TNCs, global institutions, and domestic players, including governments and domestic corporations—while the U.S. remains as a major power in the global broadcasting system in the early 21st century, and this trend is not going to fade away in the near future.

CONCLUSION

This chapter has analyzed the transformation of the global broadcasting industry amid neoliberal globalization. The global broadcasting industry has fundamentally changed under the name of liberalization and transnationalization since the mid-1980s. The broadcasting industry had been a symbol of national identity, unlike other economic and communication sectors; therefore, in many countries, television remained primarily a national phenomenon, and national governments played significant roles in protecting the broadcasting system through their legal force. The neoliberal reform, however, has shifted the global broadcasting industry from a mainly government-dominated or founder/owner-controlled sector to a profit-driven private sector. The pursuit of privatization and transnationalization of the broadcasting system allowed the inclusion of new commercial broadcasting corporations, instead of focusing on cultural identity and the public sphere. The growth of the global broadcasting industries has also been possible due in great part to the development of the new technologies of cable, satellite, and digital as vital economic sectors (Jin, 2007). The impacts of these new communication technologies along with neoliberal reform have transformed the global broadcasting landscape, from terrestrial channels to multi channels, such as satellite, IP TV, and digital TV. The rapid growth of multi channels has expedited the decreasing role of existing terrestrial channels. Although broadcast television stations have increased in number, their total viewership steadily declined. In fact, the audience share of network affiliate stations for the original major networks in the U.S. (ABC, NBC, and CBS) shrank from 64.2% in 1984 to 34.1% in 2005 (Noam, 2009, p. 64). These terrestrial channels have expanded through M&As; however, they now pursue de-convergence as in the breakup of Viacom-CBS. For broadcasting industries, M&As have been used as a very strong tool for companies to become media giants; however, the same method has suffocated their corporations, which has resulted in a change of their business strategies, from convergence to de-convergence.

In the midst of neoliberal media policy, TNCs, in both media and non-media sectors, played a significant part in the process of change. TNCs have invested an enormous amount of money in the broadcasting industry in developing countries as a form of diverse investments, including joint ventures and coproductions, but mostly through corporate integration, because it became a highly profitable sector of the global economy. Although several broadcasting corporations located in developing countries have expanded their investment, a few Western-based broadcasting companies have rapidly increased their dominance in the global broadcasting market. Through convergence strategies, these Western-based media TNCs have become media giants, and the rise of mega broadcasting corporations has allowed large companies to control both content and channels to maximize their dominant positions.

However, the global broadcasting industries have experienced a huge surge of de-convergence in the 21st century. While the convergence of the broadcasting system still remains a powerful force, de-convergence has become prevalent after the failure of several merged broadcasting corporations, such as Viacom-CBS and AOL-Time Warner, and many broadcasting mega giants around the world are pursuing a new direction. In particular, several network broadcasters have been struggling due to new technologies, such as cable, satellite, and the Internet TV. Therefore, they have to cut their size through de-convergence in order to focus on content businesses, as will be detailed in Chapter 8.

In sum, the global broadcasting industry has changed and has been influenced by sometimes cooperative and at other times conflicting relationships among the national government, domestic capitals, and transnational corporations over the past two decades. While TNCs have invested in and expanded their media areas, mainly with television, governments around the world played pivotal roles in the transformation of the broadcasting industry through deregulation measures.

4 Transnationalization of the Advertising Industries

Advertising has been a major revenue resource for media industries. From broadcasting to newspaper to social media, advertising is always significant for their operation. Advertising and old media, including broadcasting, have depended on each other. With the development of the Internet, however, the situation has gradually shifted because advertisers have changed their major outlets from traditional media to the Internet. The rapid growth of new media, including social media, such as Facebook and Twitter, has further shifted the traditional relationship between media and advertising. While old media have experienced a substantial decrease in advertising revenue, social network sites (SNSs) and user-generated content (UGC) outlets, including YouTube, have enjoyed increasing revenues from advertising in the early 21st century.

With the rapid growth of the advertising market in tandem with social media, as well as the transformation of global capitals, the advertising agency business itself has rapidly consolidated over the last two decades. Both small local advertising agencies and global mega advertising agencies have changed hands in order to survive and prosper in the midst of neoliberal globalization. Advertising agencies have also expanded their predominant business models, from traditional advertising to marketing and promotion, in order to adjust to the transformation of the global markets. Since advertising has been significant for both old media and new media, it is crucial to understand the recent transformation of the advertising industry as a main part of the global communication system.

This chapter mainly investigates and analyzes the external/structural consequences of advertising, although the critique of advertising comes from a range of perspectives and viewpoints. This view broadly takes a political-economy perspective and centers on the institution of advertising as an instrument of corporate or monopoly capitalism (Dyer, 1984). It examines the historical development of the global advertising industry primarily through mergers and acquisitions between the early 1980s and 2012. Its essence lies in an empirical analysis of the structural change and dynamic of the advertising industry. It especially explores the role of U.S. advertising corporations to determine whether the U.S. has taken a pivotal role in the global M&A market, as in the case of the cultural market. This leads me to raise the fundamental question of whether advertising corporations in non-Western countries have expanded their influence on the global markets so that they can diminish an asymmetrical power relation between the West and the East.

THE MAJOR ROLE OF ADVERTISING IN THE MEDIA SYSTEM

Advertising is a key component of the rapid growth of the global communication system because advertising enables much of the production of content occurring in other commercial media (Deuze, 2007, p. 114). Global advertisers provide commercial support to the burgeoning global communication industry; therefore, the global expansion of television and other media could not have been possible without the support of advertising, central to a commercial media culture (Thussu, 2006, p. 126). The transition to consumerism is also closely related to the development of the communication industries because communication industries are essential tools in promoting it. Advertising persuades consumers to buy products available in the market, and television, the Internet, and social media are attractive vehicles for advertising.

Advertising and television broadcasting, in particular, have relied on each other; television delivers audiences to advertisers and by extension to corporations, which in turn provide the bulk of television revenues. With the proliferation of new private advertising-supported TV channels and the tremendous growth of commercial television in Europe, Latin America, and Asia, international advertising expenditures have greatly increased over the last two decades. World advertising expenditures in constant dollars was only valued at $66.1 billion at current prices in 1980, but increased to $182.5 billion in 1990 and soared to $290.1 billion in 1999.[1] According to ZenithOptimedia's annual December forecast of global advertising expenditures, world advertising expenditures in 2002 were $311.7 billion in major media, including TV, print, radio, cinema, and the Internet (ZenithOptimedia, 2002).[2] This figure soared up to $450 billion in 2010, a 44% increase over the past eight years (ZenithOptimedia, 2011). Global ad expenditure, again, rose to $466 billion in 2011 and was expected to be $498 billion in 2012 (ZenithOptimedia, 2012). The global ad spending had sharply declined between 2007 and 2009 due to economic recession; however, it has gained momentum and bounced back in recent years.

International advertising, as in the case of the film and television markets, has been dominated by the U.S., which is not the only the largest market but also the home of the largest transnational companies and advertising agencies. The U.S. was the largest advertising market in 2002, with $228 billion in advertising expenditures, followed by Japan (Global Insight Inc.,

2002). U.S. advertising expenditures peaked at $234 billion in 2007, but the market size shrank to $154.1 billion in 2011, due to the economic recession in recent years starting in 2007 (ZenithOptimedia, 2012).

However, advertising in the developing world has been growing faster, driven by its much faster economic growth. While it accounted for only 8% of total advertising spending in 1992 and 13% in 2002, it soared to 41% in 2010 (ZenithOptimedia, 2011; 2002). In particular, the growth of advertising in several Asian, Middle Eastern, and Latin American countries has been considerable. The largest increase has been in China, which is the third largest ad market in the world. In 2005 the Chinese advertising market was 23% of the size of Japan; yet in 2010 it was 57% and by 2013 it is projected to be 82%. Brazil, at sixth place, is close to the U.K., being 81% of the size of the U.K. in 2010, and it will be 89% in 2013 (ZenithOptimedia, 2011; Table 4.1).

Advertising has been closely related to contemporary capital economy. Several Western-based mega industries, such as the electronics, auto, oil, telecommunications, food, and clothing industries, have expanded their penetration in the global markets in order to create new profits. Advertising is indeed one of the most significant ways that the global media system is aligned to the global market economy. Advertising is conducted disproportionately by the largest firms in the world, and it is a major weapon in the struggle to establish new markets (McChesney, 2008). Global communication systems provide the main vehicle for advertising corporate products for sale, thereby facilitating corporate expansion into new nations, regions, and markets. Consumer product manufacturers, such as Procter & Gamble, General Motors, Johnson & Johnson, and Coca-Cola, as the largest advertisers, seek appropriate media to introduce and sell their products worldwide. The top twenty largest manufacturers, including P&G, Unilever, L'Oréal, General Motors, Nestlé, Toyota, and Coca-Cola spent $61.31 billion on adver-

Table 4.1 Top 10 Ad Markets in 2010 (millions of dollars)

1	U.S.	151.6
2	Japan	46.1
3	China	26.1
4	Germany	23.7
5	U.K.	18.1
6	Brazil	14.7
7	France	12.5
8	Australia	11.2
9	Italy	10.2
10	Canada	10.0

Source: ZenithOptimedia. (2011). Global Ad Expenditure, Press Release, October 3.

tising, accounting for 13% of the global advertising market expenditure of $450 billion in 2010, up from 10% in 2001 (Dagmal, 2011; *Advertising Age*, 2002). Among these, P&G spent as much as $11.4 billion, followed by Unilever ($6.62 billion) and L'Oréal ($4.98 billion). Several auto corporations, such as General Motors and Volkswagen, also spent ample amounts of money for advertising in several different media (Table 4.2). The role of a small number of mega transnational companies has become increasingly greater in the global communication and advertising industries.

Global advertising expenditure by medium has shown a dramatic change with the rapid growth of the Internet and, now, social media. As usual, the main contributor to global ad growth is television, accounting for 39.8% of global ad revenues of the media. Regardless of the rapid growth of the Internet as a new ad medium, television is expected to continue to grow and maintain its leading position, because the amount of time viewers spend watching television continues to increase (ZenithOptimedia, 2011). When Samsung launched its smartphone, Galaxy S III, in the U.S. in June 2012, a major component of the ad campaign was the use of movie theaters, as Samsung wanted to be where consumers typically frequent in the summertime and made sure Samsung was there in an innovative and never been done before way. Samsung developed a 3-D game format for this particular line of ads in some 2,000 American theaters (*Advertising Age*, 2012). While the significance of social media as a major outlet is increasing, the convergence of new media, in this case the smartphone, and old media, in this case theaters, has become a trend in the advertising market. Newspapers and magazines have been declining since 2007, and this trend will continue in the near future. In fact, advertising expenditure in newspapers has declined from 23.2% in 2009 to 20.2% in 2011, and it is expected to be 17.9% in 2013.

The Internet has become the third largest medium for advertising spending, only behind television and newspaper, and the Internet is growing much faster than any other media, at an average of 14.6% per year between 2010 and 2013. This means that in 2013 the Internet will surpass newspapers in global ad spending. What is significant with the Internet is the rapid growth of social network sites, such as Facebook, MySpace, and Twitter, as the most significant part of the industry. The connection of SNSs to capitalism is especially significant. SNS users provide their daily activities as free labor to network owners, and thereafter, to advertisers, and their activities are primarily being watched and counted and eventually are appropriated by large corporations and advertising agencies (Jin, forthcoming). As the number of SNS users has soared, advertisers, including corporations and advertising agencies, have focused more on SNSs as alternative advertising media. Advertisers were expected to spend $3.3 billion in 2010, up from $2.2 billion in 2009, to advertise on SNSs worldwide, with $1.7 billion in spending in the U.S. alone in 2010 (eMarketer, 2010; *Adweek*, 2009). Not surprisingly, about $1.2 billion was projected to go to Facebook in the U.S., and, of course, these SNSs have profited from foreign countries as

Table 4.2 Top 20 Global Ad Marketers in 2010 (billions of dollars)

1	Procter & Gamble Co.	11.40
2	Unilever	6.62
3	L'Oréal	4.98
4	General Motors	3.59
5	Nestlé	3.19
6	Toyota	2.86
7	Coca-Cola	2.46
8	Reckitt Benckiser	2.43
9	Kraft Foods	2.34
10	Johnson & Johnson	2.32
11	McDonald's	2.32
12	Volkswagen	2.24
13	Ford	2.14
14	Sony	2.04
15	Mars Inc.	1.99
16	FIAT	1.79
17	Pfizer	1.72
18	Danone Group	1.68
19	GlaxoSmithKline	1.61
20	PepsiCo	1.59

Source: Dagmal, K. (2011). 100 Global Marketers. *Advertising Age 82*(43): 6–7.

well (eMarketer, 2010). It is certain that both advertisers and advertising agencies will target social media as major outlets in the near future as social media grows exponentially.

CONVERGENCE OF THE ADVERTISING INDUSTRY

On an institutional level, two major developments run throughout the advertising industry: business fragmentation through spin-off and split-off and concentration through M&As. These trends must be seen as co-constituent since many professionals in advertising start their own small-scale enterprises, while, at the same time, the existing conglomerates keep expanding their reach by acquiring certain successful or specialized smaller companies in the field. In particular, large investment networks like Publicis, Omnicom, or Harvas have been gobbling up agencies worldwide at an accelerated pace since the 1980s in an attempt to become transnational full-service companies—business that offer clients and advertisers a wide range of products and services, including advertising, marketing, and public relations (PR) (Deuze, 2007, pp. 114–115). The advertising industry, like most other media industries, is highly concentrated in the hands of a few transnational advertising corporations. Although there are hundreds of thousands of accredited advertising agencies, the majority of the total business goes through the top 100 agencies. In addition, in most capitalist countries, the majority of the large agencies operating in those countries are either subsidiaries or affiliates of Western-based TNCs.

In fact, the advertising agency business itself has rapidly grown and consolidated on a global basis over the last two decades, in part to better deal with the globalization of product markets, and also to better address the plethora of commercial media emerging to serve advertisers (McChesney, 2008). Above all, the leading global advertising agencies tend to be smaller than leading media firms, whose annual sales are as much as 10 times greater (*Business Week*, 1996). The expansion of advertising internationally in the 1980s fuelled the creation of mega advertising agencies, as the result of widespread mergers and acquisitions. In the 1980s, there was a major wave of M&A among advertising agencies, especially in the U.S., and the trend toward consolidation in the 1990s has expanded (Herman & McChesney, 1997, p. 59).

Until the mid-1980s, the function of the traditional advertising agency consisted mainly of planning and creating advertising campaigns and choosing the appropriate media in which to place ads for their clients. The development of marketing strategies and the positioning of products in a given market were functions of the marketing department of a corporation or an independent marketing organization. Thus, until the 1980s, these two functions (advertising and marketing) were in general handled separately. By the mid-1980s, however, the mega agency increasingly undertook the dual role of marketing and advertising (Mahamdi, 1992, pp. 160–161). The recession of the 1970s and early 1980s forced many ad agencies to diversify into other sectors, including marketing and research services to gain additional revenue. Large consumer companies began shifting budgets from long to shorter-term marketing strategies in order to show attractive earnings to shareholders. This led some agency managers to acquire firms specializing in direct marketing, sales promotion, and health services to fill gaps in agency offering (Ducoffe & Smith, 1994). Horizontal integration involving medium or large agencies acquiring smaller counterparts characterized the 1980s; however, the merger and acquisition trend continued with the creation of superagencies in the 1990s. It appeared to some that the only motivation behind such activity was size (Ducoffe & Smith, 1994).

The overall and international mergers and acquisitions in the advertising industry increased rapidly between the 1980s and the early 21st century. During the same period, an overall 5,629 mergers and acquisitions were completed. The total transaction values were $170 billion. There were only eight M&As in advertising agencies in 1982; however, they have increased considerably beginning in 1988 (160 deals). In particular, the number of annual deals completed peaked between 1997 and 2001. In the midst of the rapid growth of neoliberal transnationalization, the number of deals had increased from 241 in 1996 to 366 in 1997, and as many as 475 deals in 2000. The number of deals in the advertising industry, however, has declined

Table 4.3 M&As in the Advertising Industry (millions of dollars)

Year	Number of Deals	Transaction Values (millions of dollars)
1982	8	575.13
1983	15	143.80
1984	36	1,091.21
1985	15	1,401.41
1986	51	3,714.61
1987	51	3,511.46
1988	160	2,226.40
1989	199	5,742.47
1990	159	2,487.78
1991	191	410.72
1992	163	1,451.13
1993	135	1,169.46
1994	182	1,868.65
1995	227	1,987.37
1996	241	4,193.28
1997	366	19,153.45
1998	432	12,650.42
1999	474	10,007.08
2000	475	25,860.80
2001	381	9,069.15
2002	250	5,007.79
2003	167	1,211.93
2004	240	3,809.64
2005	200	7,436.55
2006	223	3,357.81
2007	259	16,948.77
2008	189	20,181.43
2009	140	3,387.87
Total	5,629	170,057.60

since 2001 due to two consecutive global recessions. The number of deals recorded was 167 in 2003 and 140 in 2009, respectively (Table 4.3). While other communication industries have also been hit by the two consecutive economic recessions in the early 21st century, advertising has been affected the most due to its close relationship with economy. The number of deals in 2009 was recorded at only about 29.5% of that of 2000, and it certainly proves that the advertising industry has not been active in the global deal market in very recent years. M&As have been a common growth strategy for advertising agencies over the past several decades. One after another, advertising agencies have acquired other ad agencies to become mega ad agencies.

Various goals for M&As are often suggested, including profit maximization, risk avoidance, enrichment of senior management, or simply the desire to form huge corporations. Smaller agencies may sell or merge due to cash flow problems, a lack of capital to make essential improvements, a loss of importance accounts, the desire to attract top personnel, or superior financial capabilities on the part of rivals in gaining accounts (Ducoffe & Smith, 1994).

Among all mergers, the U.S. acquired as many as 2,167 ad agencies (38.5%), and non-U.S. agencies acquired 3,462 companies (61.5%). This data proves that advertising agencies themselves have rapidly grown and reshaped themselves on a global basis. It also shows that M&As in advertising agencies occurred very actively in non-U.S. regions. Advertising agencies have become one of the fastest communication industries invested in by non-U.S. based transnational companies beginning in the mid-1980s. However, one needs to be very careful about this data, because when we include only a few other Western countries, these countries are still major players. In fact, the top seven Western countries, including the U.S., the U.K. (1,032), France (410), Japan (281), Australia, Germany, and Canada, consisted of 79.7% of the deal market. These Western countries have been active in acquiring advertising agencies both globally and locally.

Meanwhile, cross-border M&As between two countries suggest similar interpretations. During the period 1982–2009, there were as many as 1,582 cross-border deals (28%). As with the overall transactions, the U.S. acquired the largest number of ad agencies (600 companies) from other countries, followed by the U.K., France, and Canada. Including Australia, Japan, and Germany, these seven big countries accounted for 82.4% of cross-border deals (Figure 4.1), which is even higher than overall M&As. As far as the cross-border deal is concerned, a few Western countries have continuously dominated the global deal market. Although several emerging markets, such as China, Brazil, Mexico, and Korea have increased their influences as capitalists in the global market, their roles are not comparable to several Western countries. In fact, during the same period, China acquired 71 foreign ad agencies, followed by Brazil (31), Korea (20), and Mexico (4). China was able to buy only one foreign agency until the late 1990s, and the country has acquired more than 10 foreign agencies in recent years; however, it is

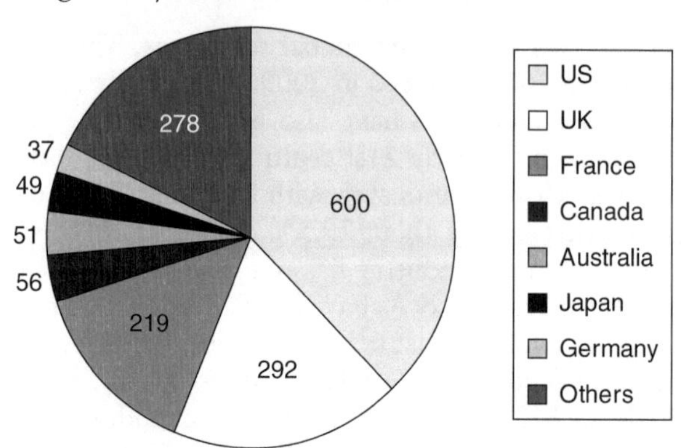

Figure 4.1 Cross-border Deals in the Advertising Industry

still far behind other Western countries. The advertising market might have expanded as a result of neoliberal globalization, but the major advertising agencies have not changed much in the past several decades.

CONCENTRATION OF ADVERTISING AGENCIES IN THE GLOBAL MARKETS

Global ad agencies have greatly expanded their global dominance through aggressive M&As, as several mega ad agencies, such as WPP (157 deals), Omnicom (105), Publicis Groupe (97), and Interpublic (90), have acquired many ad agencies both nationally and globally. For example, WPP, the largest ad agency in the world, agreed to buy Commarco Holding, a German agency group, and it completed 25 small and midsize acquisitions in 2000, including the Canadian agency Taxi, I-Behavior, and Marketing Direct Inc., a U.S. database marketing firm (Bradley, 2011). These agencies also hope that buying the in-house agency of a Korean marketer with a big global ad budget would give them the inside track on those ad dollars. LG Group, for instance, spent about $200 million outside Korea, divided among many agencies in 2002 (Lee, 2002). WPP acquired LG Ad, the second largest Korean ad agency, as part of LG Group, at $42 million (31% stakes) in 2003, although the LG group retook LG Ad and changed its name to HS Ad as an in-house agency in 2008. In 2003, WPP bought another big local agency, *Kumgang*, the third largest agency in Korea, and formed *Kumgang* Ogilvy. As such, one of the most emergent markets has been targeted by transnational advertising agencies, and they have shaped and reshaped the structure of the Korean advertising industry (Ko & Cha, 2009). As Table 4.4 shows, transnational advertising agencies from several Western countries are major players in the Korean market.

Korea is one of the countries that transnational ad agencies are targeting. Omnicom, the second largest ad agency in the world, took majority ownership

Table 4.4 Major Foreign Ownership of Korean Advertising Agencies in 2011

Name of Agency	Foreign Agency	Country	Ownership by Foreign Agency
Grey Worldwide Korea	Grey Global Group	U.S.	100%
Kumgang Ogilvy	WPP	U.K.	100%
Dentsu Media Korea	Dentsu	Japan	67%
DDB Korea	DDB Worldwide	U.S.	94%
BBDO Korea	Omnicom	U.S.	70%
Welcomm	Publicis	France	majority
Euro Next	Havas	France	80%
TBWA Korea	Omnicom	U.S.	100%
JWT Adventure	WPP	U.K.	80%

Source: "Research in Advertising Industries." *Trends in Advertising World*, March 2011, p. 29.

of Clemenger Group, an Australian agency company, and purchased Communispace, a digital market research services firm, and bought Sales Power, an in-store promotion company in China. Publicis also acquired PR firms Eastwei Relations (now Eastwei MSL) in China and Interactive Communications Ltd. (now ICL MSL) in Taiwan.

As China has vaulted from emerging market status to the point of being the third largest national advertising market after the U.S. and Japan, alongside its market liberalization, several Western-based mega advertising agencies have rapidly invested in the Chinese market, although joint ventures with domestic advertising corporations are preferred to M&As (Sinclair, 2008). Foreign advertising agencies, as in many media sectors, do not want to make conflicts with the Chinese government and corporations. It is interesting that while this restriction was lifted in January 2006, under China's terms of entry to the WTO, the joint venture model continues to be favored by the many global agency networks that have been attracted to China. For example, although one of the first to enter in its own right under the new regime was the U.K.-based agency network, Bartle Bogle Hegarty, 49% owned by the French-based Publicis Groupe, Publicis also announced a new joint venture later that same year (Publicis Groupe, 2006, cited in Sinclair, 2008; Trombly, 2006, cited in Sinclair, 2008, p. 78). No doubt more foreign agencies would be attracted to China by this recent liberalization of foreign ownership in the advertising business, particularly since it parallels a similar lifting of restrictions in other industries, such as banking and tourism, which are clients of advertising (Sinclair, 2008). In addition, the 2008 Olympics in Beijing and the 2010 World Expo in Shanghai were added attractions, against an overarching background of exceptional economic growth (Trombly, 2006, cited in Sinclair, 2008). However, for the present, apart from several Chinese agencies, the list of leading agencies in China is characterized by joint ventures, notably Beijing Dentsu, Shanghai Leo Burnett, Saatchi & Saatchi Great Wall, McCann Erickson Guangming, and Shanghai Advertising (partner to both Ogilvy & Mather and Hakuhodo) (Sinclair, 2008).

Several mega global ad agencies have indeed rapidly increased their investments in Asia and other emerging markets, primarily because the ad markets are swiftly growing in these areas. Asia-Pacific is vital to the future growth of the global ad sector, and global ad agencies are incredibly committed to growing in Asia, recognizing that Asia-Pacific is the growth engine for most marketing communication over the course of the next 20 years. The role of global ad agencies has rapidly grown since the late 1990s and has been expedited in the 21st century. In 1996, foreign agencies accounted for merely 3.3% of the Korean market; however, this soared to 37.5% in 2000 and 57% in 2004 (Yoo, 2005). Due to the competitiveness of domestic ad agencies, including *Cheil* Communications, local-based ad agencies have taken over the majority share in the Korean market; however, foreign players, such as WWP and Dentsu, are still playing key roles.

Interestingly, although Japan's advertising market is the second largest in the world, the presence of global networks of agency brands, such as Omnicom, WPP and Interpublic, is very limited. Unlike other Asian countries, their total share of the Japanese market is negligible. Instead, the top three Japanese agencies—Dentsu, Hakuhodo DY Holdings, and ADK—almost form an oligopoly, occupying about half of the market, and these ad agencies have rapidly expanded their presence in other Asian countries, competing with other global ad agencies (Kawashima, 2009). As noted in Chapter 1, Japan has been very unique due to its advanced status in the media industry. As in the newspaper and broadcasting industries, Japan needs to be considered as part of the West, and one can certainly understand that Japan has played a major role in the global advertising market as a capitalist country.

Meanwhile, with the rapid growth of global advertising markets, the role of U.S. advertising agencies has declined, although the U.S. is still a major power in global advertising. U.S. advertising agencies were the top 10 worldwide agencies until the early 1980s; however, the U.S. share of the global advertising business appears to have declined, while those of Japanese and European countries appear to have increased. In 1980, for example, all of the world's top 10 agencies were U.S. agencies, except for Dentsu of Japan. In 1989, however, due to the consolidation of large agencies, only five transnational advertising agencies were U.S. agencies, after J. Walter Thompson (1987) and Ogilvy & Mather (1989) were sold to the British WPP Group, one of the largest marketing and communication service companies in the world (Oram & Tait, 1989, p. 1). The WPP group includes agencies and bureaus offering services in advertising, media investment management, PR, and all kinds of marketing communication, employing 153,000 people in more than 2,400 offices in 107 countries (WPP, 2012). Non-U.S. based advertising agencies grew further in recent years. In 2002, 7 of the 10 largest advertising agencies in the world were non-U.S. companies, including three Japanese agencies (Dentsu, Hakuhodo, and *Asatsu-DK*), two British (WPP and Cordiant Communications), and two French agencies (*Publicis Groupe* and Havas) (Endicott, 2003, p. 3). Omnicom, Interpublic, and Grey Global were three U.S. advertising agencies in the top 10 global ad agencies in the same year. This trend has continued until recent years. In 2010, among the top 10 global ad agencies, only three agencies (Omnicom, Interpublic, and Acxiom) were owned by the U.S.

There are some challenges to the U.S. dominance of the global advertising industry, mainly from U.K., French, and Japanese agencies; however, most developing countries still cannot challenge the U.S. agencies, so only a few countries' agencies have dominated the global advertising market in recent years. In fact, in 2010, the top 20 global agencies consisted of 12.5% ($56.3 billion) of the global ad marketing, which means that they are major players in this market. Among the top 20 global ad agencies, 90% are owned by

Table 4.5 Top 20 Global Ad Agencies in 2010

	Corporations	**Locations**	**Revenues (millions of dollars)**
1	WPP Group	U.K.	14,416
2	Omnicom Group	U.S.	12,543
3	Publicis Groupe	France	7,175
4	Interpublic Groupe	U.S.	6,532
5	Dentsu Inc.	Japan	3,600
6	Aegis	U.K.	2,257
7	Havas	France	2,069
8	Hakuhodo DY Holdings	Japan	1,674
9	Acxiom	U.S.	785
10	MDC Partners	Japan/U.S.	698
11	Alliance Data Systems Corp.'s Epsilon	U.S.	613
12	Groupe Aeroplan's Carlson Marketing	U.S.	593
13	Daniel J. Edelman	U.S.	544
14	Sapient Corp.'s Sapientnitro	U.S.	515
15	ASATSU-DK	Japan	484
16	Media Consulta	Germany	408
17	Cheil Worldwide	Korea	386
18	IBM Corp.'s IMB Interactive	U.S.	368
19	Grupo ABC	Brazil	362
20	Photon Group	Australia	332

Source: Bradley, J. (2011). Agency 2011 Report. *Advertising Age 82*(17): 24–41.

Western countries, including the U.S. (Omnicom Group, Interpublic Group, and Acxiom), the U.K. (WPP and Aegis), France (*Publics Groupe* and Harvas), and Germany (Media Consulta). Only one of them is owned by Korea (*Cheil* Worldwide) and one by Brazil (Grupo ABC) (Table 4.5). While we cannot say that U.S.-based advertising agencies dominate the global advertising industry, it is certain that only a handful of Western countries have dominated the global ad market.

Regardless of their different nation base, however, the mega advertising agencies provide a range of services, which includes market research and total campaign development and execution on all levels—local, regional, national, and global. Naturally, the role and market portion of the transnational advertising agencies are growing greatly. For example, the big three—the U.S. Omnicom Group, the British WPP Group, and the U.S. Interpublic—had combined revenues of $8 billion in 1995, which jumped to $19.5 billion in constant dollar terms in 2002, accounting for 60.8% of the total revenue of $32.07 billion of the world's top 25 advertising agencies (*Advertising Age*, 2003, p. 4). In 2010, WPP, Omnicom, and Publicis Groupe had combined revenues of $34.1 billion, accounting for 60.5% of the total revenue of $56.3 billion of the world's top 20 advertising agencies.

CRITICAL INTERPRETATION OF THE CONVERGENCE OF THE ADVERTISING INDUSTRY

The dominant position of Western-based agencies in advertising in many countries is apparently a post–World War II phenomenon, and is closely associated with the expansion of American capital and the emergence of transnational corporations as the dominant producers in industries such as processed food, soft drinks, pharmaceuticals, cars, household appliances, soaps, cosmetics, energy, and so on (Bonney, 1984, pp. 35–36). Under capitalist relations of production, the decisive factor in the success or failure of corporations, and industries as a whole, is profitability, and that in turn depends upon growth—increases in market share, expansion of existing markets, and the development of new markets. Through the long boom of the 1950s and 1960s, many large corporations based in the U.S. (and in Western Europe and Japan) established themselves in numerous foreign countries to exploit local markets and, in the case of the Third World, to take advantage of cheap labor. In doing so, they needed to export not only capital and expertise but also marketing and advertising skills and techniques (Bonney, 1984).

Advertising agencies began to follow their large clients abroad, establishing local agencies in countries where their clients were operating, because advertising agencies were usually pulled abroad by the prior expansion of their clients (Bonney, 1984; United Nations, 1979, p. 2). A salient feature of the advertising industry is the important role played by transnational advertising firms, especially those based in the U.S. (United Nations, 1979, p. 2). Whereas only 50 foreign branches of U.S. advertising agencies were established in the period 1915–1959, U.S.-based agencies opened or acquired 210 foreign branches between 1960 and 1971. In most cases, their major clients were transnational corporations. For example, though McCann Erickson's Sydney office had 39 clients in 1981, over half the business came from eight transnational clients: Nestlé, Coca-Cola, Johnson & Johnson, Philips, Levis, Goodyear, Gilbeys, and Amatil (Bonney, 1984).

Advertising is crucial for the growth of the communication industries; therefore, the convergence of advertising agencies as a form of M&As is very significant to understanding not only the advertising market but also the communication market. Due to its influence on the communication industries, there are some controversies about the major reasons for M&As in the advertising industry. Some supporters certainly argue that increasing size resulting from M&As many permit agencies to realize certain economies of scale, such as increased negotiating strength with media suppliers. In addition, large agencies can offer a broader range of services to their clients, which can be efficiently coordinated to maximize their marketplace synergy. M&As can also help agencies expand into new domestic and international markets (Ducoffe & Smith, 1994). In fact, the wave of global consolidation among advertising agencies is far from over. Global consolidation is encouraged because the larger an ad agency, the more leverage it has getting favorable terms for its clients with global commercial media (McChesney, 2008).

More interestingly, global market integration and the growth of multinational companies has led to a surge of products and services sold on a multi-territory basis. This trend is gathering pace in the services sector because of the growth of cross-border mergers and takeovers worldwide. The number of global brands is growing every year. The names of products that used to differ from territory to territory are being standardized. These brands will always use local media for efficiency purposes and reach consumers in great numbers. But media buying agencies have realized that cross-border advertising allows them to develop these brands homogenously across a region and achieve consistent brand image and positioning (Chalaby, 2008, pp. 143–144). As a result, the number of global advertisers in the world has steadily climbed throughout the 1990s and the early 21st century. The formation of global media conglomerates prompted the advertising industry to consolidate its media buying capacity. The advertising industry faced an increasingly complex media environment, due notably to the multiplication of content platforms and audience fragmentation (Chalaby, 2008).

Critics of M&As, however, maintain that results will be less favorable. Several scholars were concerned about the dominant positions of a few mega ad agencies, primarily because of their control over capital. Since the early development of advertising, the market share of the transnational agencies' operations outside the country of their origin had been dominated by some of the top Western agencies (United Nations, 1979, p. 9). As Baran and Sweezy

(1968) argued, in a capitalism dominated by large corporations operating in oligopolistic markets, advertising especially becomes a necessary, competitive weapon. In other words, in an economic system in which competition is fierce and relentless and in which the fewness of the rivals rules out price-cutting, advertising becomes to an ever increasing extent the principal weapon of the competitive struggle (Baran & Sweezy, 1968). Accordingly, the advertising business has grown astronomically, with its expansion and success being continually promoted by the growing monopolization of the economy and by the effectiveness of the media which have been pressed into its service—especially radio, and now, above all, television (Baran & Sweezy, 1968, pp. 115–117). What has actually happened is that advertising has turned into an indispensable tool for a large sector of corporate business. It has become an integral part of corporations' profit maximization policy and serves at the same time as a formidable wall protecting monopolistic positions. In other words, advertising constitutes as much an integral part of the system as the giant corporation itself.

Advertising must also be seen as integral to consumer capitalism. As noted, the advertising agency business itself has rapidly consolidated on a global basis beginning in the mid-1980s. It can be argued that the creation of the mega advertising agency is part of the process of globalization whereby transnational corporations, media conglomerates, and transnational advertising agencies are brought together to extend the close relationship between production and consumption on a global scale, a necessary condition for the expanded reproduction of consumerism. This partnership between business, communication, and advertising is not new, but the expansion of this partnership into a global convergence of interests is the driving force behind the emergence of a global consumerism. As Dyer (1984) already pointed out, modern advertising arose because of the shift from competitive to monopoly capitalism at the turn of the century and, because of the inherent irrationality of the capitalist system of production, the market had to be controlled. In addition, the 'sponsorship' of commercial TV, newspapers, and magazines by advertisers means that what ought to be public communication systems are in general subordinated to the needs of advertisers both in terms of style and content.

CONCLUSION

The focus of this chapter has been on the dimensions of transformation of the global advertising industry. The global advertising market and industry has dramatically changed since the mid-1980s, and these changes intensified in the 1990s. Since the 1980s, M&As as commonly employed strategies for advertising agency growth have continued more or less uninterrupted other than during or owing to recent global recessions in the 21st century. The changing political-economic environment in the world caused the rapid transformation of the global advertising industry. The introduction of neoliberal economic policies has rapidly changed the global advertising system. Adoption of deregulation and the liberalization of advertising systems beginning in the mid-1980s in the U.S. and the U.K. allowed for the inclusion of new commercial advertising companies in the world.

A few Western agencies today remain the dominant forces of global advertising. Several mega ad agencies in the U.S., the U.K., and France have found much faster growth rates overseas than in the domestic market, and, as a result, global billings continue to represent a more significant share of the total revenues of large Western agencies. Transnational advertising corporations have played significant roles in the process of change with international organizations and the U.S. government. These international forces served as driving forces to the neoliberal reform in the communication sector. With deregulation in each government, TNCs have invested an enormous amount of money in the communication industry in developed countries because it became a highly profitable sector of the world economy, which resulted in development at ever increasing rates of growth in the advertising industry.

Moreover, the role of transnational capitals has grown in the advertising sector with new information technologies, including wireless telephone, the Internet, and broadband, since the early 1990s. Indeed, while television was the primary entertainment and information medium in the 1990s, telecommunications and information technologies in the late 20th century and early 21st century have expanded the scope of international communication. In the 21st century, the advertising industry is evolving to include advertising and media, public relations, branding, marketing research, and corporate communication, used not just by commercial companies, but also by government and nongovernmental organizations, to brand their versions of truth to an increasingly media-savvy and fragmenting audience (Thussu, 2006, p. 129).

The global advertising industry cannot be completely free from the concern that globalization and transnational forces drive the industry to a more profit-driven commercial-dominant environment (Ko & Cha, 2009). Although several emerging countries have challenged the dominant power of transnational agencies in international markets, Western-based mega agencies have expanded their penetration in the global market. Local forces have come to coexist with Western mega giants through both competition and cooperation, and global forces have taken pivotal roles. While other media industries, such as television and newspaper, have seen similar trends, advertising has more closely related to the global capital economy than these media institutions, and with the transnationalization of Western corporations, Western advertising agencies have rapidly invested in non-Western markets as a form of M&As and joint ventures in order to utilize local tastes for local consumers.

5 Convergence of the Movie Industries

Media integration has been among the most distinctive in the film industries in the past several decades. From Hollywood majors, such as Warner Brothers and Walt Disney, to small film exhibition firms in developing countries, including Korea, Mexico, and Argentina, many film corporations have expanded their investments, both domestically and globally, and have transnationalized their businesses through corporate integration. Due to their importance as both cultural symbols and money-making businesses, many media corporations around the world have increased their capital involvement in the film industries.

Media integration of cultural industries as a form of M&A is not new; however, the degree of the integration of the film sector has been prominent because consolidation through industry alliances and mergers has become a significant corporate policy in expanding the influence of these companies. The integration of film corporations since the mid-1980s has especially been prevalent due to the increasing frequency of integration and the magnitude of integration in terms of transaction monies. With changing media environments, film corporations in many countries have rapidly transformed their structures through corporate integration, which has facilitated the growth of mega film giants in Western countries, including Hollywood majors (Lorenzen, 2008; Kutz, 2007; Miller et al., 2005; Bagdikian, 2004; Wasko, 2003; Collette & Litman, 1997).

This chapter intends to examine the historical development of the global film industries primarily though horizontal and vertical integration between the late 20th and the early 21st century. Its essence lies in an empirical analysis of the structural change and dynamics of the film industries. It maps out the role of U.S. film corporations—considered the key players in the global film market through Hollywood movies—to determine whether the U.S. has taken a primary role in the global deal market. It eventually debates whether film corporations in developing countries have expanded their influence on the global market, and therefore, whether they are able to reduce an asymmetrical power relation between Western and non-Western regions.

MEDIA CONVERGENCE IN THE GLOBAL FILM INDUSTRIES

There are several different forms of media integration in the film sector, including joint ventures and conglomeration. Among these, vertical and horizontal integrations are the most significant and active in expanding the scale of film corporations, although they are not mutually exclusive. The development of vertical integration has been contentious since the earliest days of the cinema in the 20th century. As explained in Chapter 1, since vertical integration—referring to the merger or acquisition of companies at different levels of production, distribution, and exhibition—makes it possible to secure resources and to directly control product specification, many film corporations have pursued vertical integration and have become bigger integrated film corporations (Sunada, 2010; Fu, 2009; Blackstone & Bowman, 1999).

Within the discourse of media integration in the film sector, horizontal integration, which is the combination of two or more companies across the same level of production and distribution, is also crucial for corporations due to scale economies and an increase in market power in media industries (McChesney, 2008; Noam, 2006; Thierer, 2005). As film corporations obtain a greater share of the market, this permits them to have lower overhead and to have more bargaining power with suppliers, while gaining more control over the prices they can charge for their products (McChesney, 1999, p. 16; *Financial Times*, 1998). Several scholars (Blair, 2001; Robins, 1993; Faulkner & Anderson, 1987; Storper & Christopherson, 1987) especially discussed the history of early horizontal integration in the production of films because production was the primary sector in the film industries.

Two primary drivers that have expedited vertical and horizontal integration in the film industries are neoliberal globalization and the increasing role of major film capitalists. These two elements are intermingled in a complicated manner, which results in the concentration of ownership in a few hands located in Western countries. Neoliberal globalization is characterized by interlocking features, including policies that promote liberalization, deregulation, privatization, and capital investment (Lindio-McGovern, 2007). Neoliberal globalization has intensified Western—the U.S. and Western European countries—and, in particular, Hollywood's, dominance in the global film market through the privatization of media ownership; a unified Western European market; openings in the former Soviet Union bloc and China; and the spread of satellite TV, the Web, and the DVD, combined with deregulation of national broadcasting in Europe and Latin America (Miller & Maxwell, 2006; Gomery, 2000).

Throughout the world, the vast majority of governments have introduced forced cultural liberalization measures, including a reduction in local government intervention in film production and opening the domestic film market, despite the fact that their support of domestic film industries is crucial for national cinema to prosper (Jin, 2006a). With very few exceptions that voluntarily opened their gates to Hollywood,[1] the majority of countries have had to open their cultural markets due to strong demands from a few Western countries, resulting in the rapid increase of Hollywood's influence in the global cultural market. The U.S. government and Hollywood majors acknowledge that the American motion picture and television production industries remain some of the most highly competitive around the world. As the core of a liberalized trade regime, the U.S. can press its capital advantages to maximum effect (Jihong & Kraus, 2002, p. 423). The U.S. has especially demanded several emerging economies open their cultural markets. The Hollywood majors have consistently outperformed their competitors. Other national markets and their leading roles have been extended in the global film business.

More important, the neoliberal globalization process requires maintaining the transnational capitalist class (Sklair, 2001)—meaning TNCs—whose policies and practices serve the interests of monopoly capital, and TNCs are the major instruments of neoliberal globalization (Lindio-McGovern, 2007, pp. 15–16). Backed by neoliberal globalization principles, film TNCs, including Hollywood majors, have integrated other film corporations with their vast amounts of capital. Over time, the majors, such as Disney, Time-Warner, Viacom, and News Corporation, as major capitalists, have consolidated and further integrated their operations, growing in size as a result (Coe & Johns, 2004). Starting in the mid-1990s, the film industries have especially witnessed an unprecedented series of M&As among global film corporations that have eventually facilitated the emergence of mega film giants. Major media capitalists have played a pivotal role in the global film market and certainly benefited from neoliberal globalization. That neoliberal globalization and major capitalists are connected is not surprising. Economic and cultural relationships always bear the imprint of powerful states and major capitals (Ikenberry, 2007, p. 41), and the film industry has been one of the major cultural economies for the West and, in particular, the U.S. Through the ongoing discussion, this chapter will shed light on current debates on the neoliberal transformation of the global film industries and will contribute to the development of current theories of media integration.

HISTORICAL TRANSFORMATION IN THE PRODUCTION INDUSTRY

Film corporations in both Western and non-Western countries have pursued horizontal integration as well as vertical integration mainly because they pursue scale economies and an increase in market power in media industries. Vertical integration was especially a key component of the Hollywood studio system during the 1920s–1960s, and even onwards.[2] Since film corporations are able to get a bigger share of the market through horizontal integration, it is also crucial to understand the increasing role of horizontal integration in the film sector. While convergence means the ownership of multiple content or distribution channels, so vertical integration is significant; however, the ownership change in the same category, which is horizontal integration, has been crucial due to its major role in the extension of market share within a few major film corporations (Gordon, 2003).[3] However, comprehensive empirical data on the integration of each film sector (production, distribution, and exhibition) is commonly lacking, and the effects of integration on global film product supply are not accordingly pronounced.

According to the SDC Platinum database,[4] overall, 13,415 cases of horizontal M&As in motion picture companies, including those in production, distribution, and exhibition, valued at $2,136 billion, were completed worldwide between 1982 and 2009. Compared to this, the number of M&As in the broadcasting industry was 11,062, valued at $1,997 billion, followed by advertising and newspaper during the same period. The horizontal integration of the film industries as a whole is the largest in both the number of deals and the total amount of transaction values among the media sectors. This implies that the film industries have been the most active media sectors in the global media deal market, primarily as they are considered as profit-making cultural genres.

Among film industries, production is the most dynamic sector. There were 8,487 deals, worth $1,261 billion in production. The majority of M&As in the film production industry have occurred in the 21st century. M&As completed between 2000 and 2009 accounted for 54.5% of all transactions, whereas they constituted 34.1% in the 1990s. The trend of M&As in the production sector had shown a gradual increase and stabilized until the mid-1990s; however, it has swiftly soared since the end of 1990s, mainly because several major media corporations jumped into the deal market with their massive capital as a result of massive market liberalization (Figure 5.1). For example, Vivendi SA in France acquired Seagram Co. Ltd, which included Universal Studios, in Canada for $40 billion and turned it into Vivendi Universal in 2000. In the same year, Viacom Inc. in the U.S. also acquired CBS Corp., which included Paramount Production, for $37.4 billion, although the merger had not fulfilled promised synergy effects, and the entity split into two companies in 2006.

Media integration in the film industries has further intensified despite two major economic recessions: The first took place right after the 2000–2001 recession, and the second was during the period 2007–2008. The deal market in the production industry, alongside distribution and exhibition, showed a downturn trend for a while in the midst of the global economic recession in recent years. M&As in other communication industries, including telecommunications and broadcasting, also decreased significantly around the world, after the September 11 terrorist attacks in 2001; therefore, the economic recession certainly played a role in the deal market in the communication industries.

What is different from other media industries is that the film production sector has witnessed a rapid resuscitation in the global M&A market in recent years. While other media industries, again telecommunications, broadcasting, advertising, and newspaper, have been struggling in the second economic recession during the period 2007–2008, partially as a result of the collapse of the housing market and the financial sector in many countries, the film industries, including production, distribution, and exhibition, have experienced a rapid recovery in the deal market.[5] In fact, M&As among film production companies peaked in 2000 with 569 deals, but there were 566 deals in 2009 as well (see Figure 5.1). Regardless of the huge impacts of the two major financial downturns in the communication industries, the global film industries have not become victims of the economic recessions. The economic cycle has had little or no systematic impact on the deal market in the film sector, which suggests that a rising economy does not necessarily help the market and a falling economy does not necessarily hurt it.

Meanwhile, by country comparison, the U.S. has been the largest player in the global M&A market. As a reflection of its magnitude in terms of the number of production corporations, U.S. film producers acquired as many as 3,190 corporations (37.6%), whether acquired production companies were U.S. owned or foreign owned, followed by the U.K. (976 cases), Japan (515), Canada, Germany, France, and Australia. These seven countries acquired 6,238 production firms (73.5%), and they dominated the global M&A market in the film production sector, where the U.S. has been a key player.

More important than this figure is cross-border deals, because it primarily tells who reigns supreme in the global deal market, and again, the role of the U.S. is prevalent. During the period 1982–2009, cross-border deals,

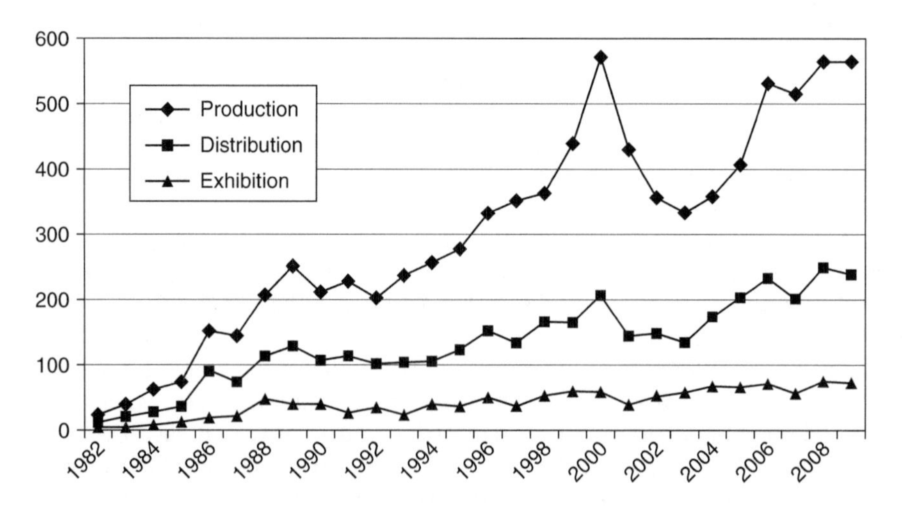

Figure 5.1 Horizontal Integration of the Global Film Industry, 1982–2009

in which target and acquirer production companies are in different countries, accounted for 1,801 cases. While several film corporations in many countries have acquired production firms in other countries, the U.S. was the country that acquired the largest number of foreign production companies. Film production firms in the U.S. acquired 440 foreign film production companies (24.4%)—meaning one country controls almost one fourth of all cross-border deals. The second largest is the U.K. (13.2%). Other major acquirer countries were Canada, the Netherlands, France, and Germany, while Japan accounted for only 65 cross-border deals (3.6%). Hollywood, as a film producer and distributor, has been considered a dominant force in the global film market, and the leading role of Hollywood in the capital market is evident.

The major benefit for this dominant position has been clear, given that Hollywood studios have successfully lowered production costs through runaway production—that is, by moving film production to countries outside the U.S. with skilled, less expensive production workers who do not belong to U.S. labor unions (Wasko & Erickson, 2008; Miller et al., 2005). Of course, hiring local film production workers brings growth to film industries outside Hollywood, so several local governments were willing to collaborate with Hollywood by forming alliances and mergers. Hollywood has created a powerful global presence as it gradually searches out inexpensive production sites in Asia, Latin America, South America, and Europe (Chung, 2007, p. 416). Because of corporate integration with foreign film producers, Hollywood has been able to further develop runaway production.

While the major role of the U.S. in the global capital market increases, non-Western countries have not exerted such power. Although some countries (for example, China, India, and Korea) are considered as emerging movie markets and are increasing their roles in the global market by providing domestically produced films, their influences as global capitalists have not been noticeable, given that they together accounted for only 2.8% of cross-border deals. Latin and South American countries, including Mexico, Brazil, and Chile, have taken no significant roles, because they acquired only a few foreign production corporations. These mid-sized economies in Asia and Latin America may have the scale and investment capacity in the global deal market; however, control over capital investment and profit remains with Western-based mega film corporations, including Hollywood majors.

NEOLIBERAL TRANSFORMATION IN THE DISTRIBUTION INDUSTRY

Distribution has become the locus of industry power due to its strategic position as the connection between production and exhibition, and this sector has been a major target from mega film companies in the global deal market.

In the film industries, distribution is wholesale and exhibition is retail; distributors lease movies to exhibitors, and organize scheduling, delivery, and collection (Miller et al., 2005); therefore, distribution has traditionally taken a pivotal role and has been considered as a major area that film corporations need to control. Since mega film corporations have integrated production and distribution due to strategic alliance, the film industry can be regarded as a producer-distributor interplay in which corporations gain market share by power selling their films to theater chains.

As the second largest sector in the film industries, there were 3,715 M&A deals in distribution, valued at $745.2 billion, which were completed worldwide between 1982 and 2009. As in the case of production, the majority of M&As in the distribution industry have occurred in the 21st century. While 54.5% of deals in the production sector occurred in the 21st century, 79% of deals in the distribution industry occurred over the past 10 years, which means that the deal market in the distribution sector has been relatively active in recent years. M&As among film distribution companies peaked in 2008 with 249 cases and in 2009 with 238 cases.

As in production, the U.S. has been the largest player in the distribution industry, and film distributors in the U.S. acquired 1,488 corporations (40%), whether the acquired distribution companies were U.S. owned or foreign owned, followed by the U.K. (313 cases), Canada, Japan, Australia, France, and Germany. These seven countries together acquired 2,756 distribution firms (74.2%), and these figures are not much different from the production industry—meaning they have dominated the global M&A market in the film distribution sector as well.

Cross-border deals in the film distribution industry accounted for 22% (818 cases) of all deals. Cross-border deals were not pronounced in the

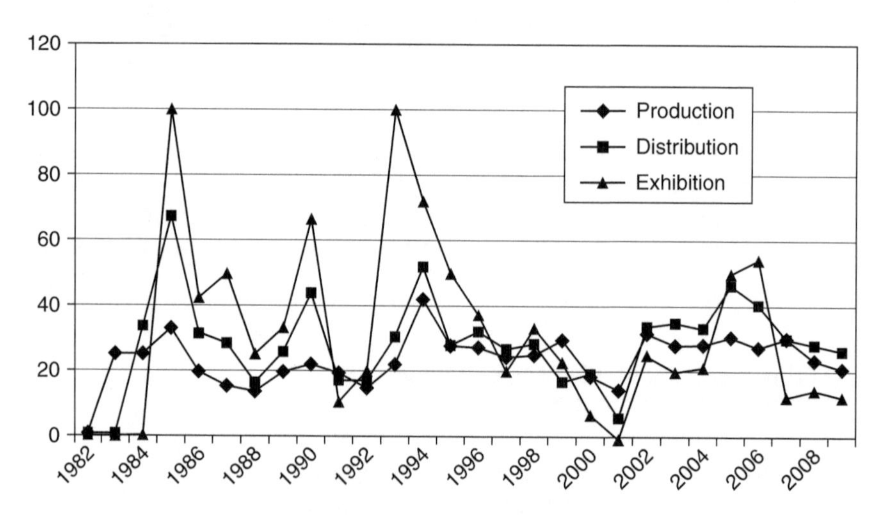

Figure 5.2 U.S. Cross-border Deals in the Movie Industries

1980s, but they have gradually increased since the mid-1990s. The U.S. was the country that acquired the largest number of film distribution firms in other countries. The U.S. acquired 283 foreign distribution companies (28.5%), followed by the U.K. and Canada. Although there are some ups and downs, particularly right after 2001 due to the consecutive economic crises, the U.S. has been a primary driver in cross-border deals in distribution, alongside production and exhibition (Figure 5.2).

In fact, the distribution industry has been dominated by U.S. firms over the last several decades because Hollywood realized early on that its dominance of the world market depended on owning the means of distribution. World distribution is controlled by the U.S. via arrangements that would be illegal domestically because of their threat to competition (Miller et al., 2005).[6] Several Hollywood majors have their own international distribution networks. The major U.S. film distributors, such as United International Pictures (UIP, for Paramount and Universal), Buena Vista (for Disney), Columbia Tri-Star (for Columbia), and 20th Century Fox (for Fox), have pursued horizontal integration in foreign counties primarily because they bring in films that have a proven box office track record in the U.S. market and are expected to appeal to global audiences (Chung, 2007). Other studios operate joint ventures that vary with the territory, sometimes with one another and sometimes with local firms or front-organization subsidiaries. Hollywood studios have sought additional control over their audiences by integrating international distribution and exhibition (Miller et al., 2005). Therefore, as Wayne Fu (2009) analyzes with the case of the Singaporean film market, the number of distributed titles and average sales shares in several local markets in which these Hollywood distributors are involved do not significantly differ across the Hollywood brands.

Consequently, several emerging markets are far behind Western countries. China, India, and Korea together account for only 2.5%, and for Latin and South American countries, the situation is even worse, so they have not taken power in the distribution sector. What is interesting is the marginal role of Japan in the global M&A market in the film industries. Japan has been one of the most significant capital investors in several communication industries, including advertising, and it has been the largest animation producer and provider; however, the role of Japan in the film sector as a capital investor has not been promising. Japan accounted for only 2.1%, even less than Hong Kong, in the distribution industry. While Western countries, in particular the U.S., have rapidly increased their capital power as global investors, several local-based media corporations and film firms have arguably extended their capital involvement in the global film market as well. However, the major players in the global capital market have not been non-Western corporations, but Western-based film corporations.

VERTICAL VERSUS HORIZONTAL INTEGRATION IN THE EXHIBITION INDUSTRY

Movie theaters have played a central role in the cultural and social lives of towns and cities in many countries. When movies were some of the primary cultural goods, people in both small towns and big cities visited theaters and enjoyed some of the best movies. However, film exhibitors around the world have experienced two opposing fates over several decades. On the one hand, film theaters have faced several challenges primarily due to an explosion of alternative outlets in which film distributors can directly sell their products, including cable television, home video and DVD, and the Internet (e.g., Netflix) since the 1970s. On the other hand, the film exhibition sector worldwide has witnessed a continuing surge with the new multiscreen theaters since the late 1980s.

Overall 1,213 deals ($129.3 billion) were completed in the exhibition industry, excluding drive-in theaters, between 1982 and 2009. M&As among film exhibition companies peaked in 2008 with 76 deals. As in production and distribution, the leading role of U.S.-owned companies in the exhibition industry has been common. U.S. corporations acquired as many as 468 theaters (38.6%), whether acquired production companies were U.S. owned or foreign owned, followed by Japan, the U.K., Australia, Canada, Italy, and Spain. These seven countries acquired 829 theaters (68.3%), and they are dominant forces in the movie exhibition sector.

The leading role of the U.S. in cross-border deals is also evident. During the period 1982–2009, there were 213 cross-border deals. The U.S. was the country that acquired the largest number of exhibition companies in other countries. Film exhibitors in the U.S. acquired 63 foreign film theaters (29.6%), followed by the U.K. (8.5%). Other major acquirer countries were Australia, Canada, and Hong Kong, while some Asian countries, such as China, India, and Korea, consisted of 4.3% of the market. This implies that, as in production and distribution, only a handful of Western countries, in particular the U.S., have played a major role in the global capital market in the exhibition industry, although the number of transactions are less than those of production and distribution.

There are several major historical factors that have influenced the exhibition industry in the U.S., which have consequently influenced other countries: the antitrust action to separate the production of films from their exhibition in the late 1940s; the advent of television in the 1970s; and the rise of VHS and DVD systems in the 1990s and the 21st century. To begin with, during the studio era (mainly until the 1940s), some studios, like MGM, RKP, Warner Brothers, and Paramount Pictures, vertically controlled theatrical outlets for first-run films, and they owned many first-run theaters (De Vany & McMillan, 2004). Their grip was so strong that the U.S. government separated the functions in a series of antitrust orders in the late 1940s. The U.S. Supreme Court concluded that controlling both the supply of films and the venue for film exhibition constituted a monopoly (this decision is known as *United States v. Paramount Pictures*), so many of the big film makers were forced to split their theater and production activities and were still subject to those orders unless they won special permission from the courts in the 1980s (American Film Institute, 2010; Gil, 2008; Harmetz, 1986). Due to this case, by 1950, over 5,000 theaters across the U.S. had closed, and Paramount alone lost some 1,395 theaters (Gil, 2008; De Vany & McMillian, 2004). In the 1980s, the *Paramount* ruling, in effect since 1950, was revisited. In a complete reversal of its original holding, the New York District Court allowed Loew's, which had restricted itself exclusively to exhibition, to produce and distribute films as long as it did not screen any of its own films (*United States v. Paramount Pictures*, 1980–1982 Trade Case. [CCH], p. 63). The court noted that much had changed in the film industry since the last time it visited the *Paramount* decision, including the introductions of television, home video, and the growth of national theater chains (Gil, 2008).

Regardless of the change in the nature of the exhibition industry with deregulation, the exhibition sector had remained subdued due to the advent of television, which directly hit the film industry in the 1970s and 1980s. Television has firmly replaced the movie theatre as the prime showcase for visual entertainment since 1985, and it has behooved Hollywood to try to control one or more important television stations (Gomery, 1986). Hollywood majors had to pay attention to this new exhibition sector, and they initiated a vertical integration spree. For example, during 1985, the U.S. television industry experienced a spate of corporate takeovers unmatched since the 1950s. In a single year, one network (ABC) was sold to Capital Cities Communications, another (NBC) taken over by General Electric, and the third (CBS) nearly toppled by cable television mogul Ted Turner (Gomery, 1986). Furthermore, in the midst of loosening antitrust regulation, MCA Inc., the parent company of Universal Studios, bought television station WOR in New Jersey for $387 million in 1986.

The studios' new surge of interest in owning theaters and TV stations stems largely from the changing economics of the business. The costs of making films have soared, and real box office smashes have become rare. With so many new movies competing for theater space in the early 1980s, theater owners had gained the upper hand, and the studios' strategy was to win control of broadcast television and theater outlets (Harmetz, 1986). While they emphasize television, both terrestrial and cable, as the major means of airing their products, film theaters themselves have also slowly increased their investments in purchasing other theaters. Film theaters have also been confronted with the surge of new technologies, including both VHS in the 1990s and DVD and digital delivery technologies in the 21st century. As technology further develops new delivery systems, including online subscription services based on the rapid growth of high-speed Internet, film theaters must compete with these new technologies.

However, the exhibition sector has witnessed substantial growth with the rise of megaplex theaters since the late 1980s. Using MPAA (Motion Picture

Association of America) data on the number of indoor screens, there were 10,335 screens in 1971 and 14,732 in 1981; however, the number of screens soared to 39,547 in 2010, mainly due to the rapid growth of multiscreen theaters (MPAA, 1986; 2010). Of course, the booming economy since the mid-1980s, alongside the growing population, has been significant for the growth of the exhibition sector in the U.S. and elsewhere because screens are being added predominantly by the construction of new complexes in or adjacent to large shopping malls (Guback, 1987).

While the number of screens has increased, ownership in the film exhibition sector has rapidly changed due to the financial difficulties of many independent theaters, and this has resulted in the concentration of ownership in the hands of a few major players. As of December 2010, the top four chains (Regal Entertainment, AMC Entertainment, Cinemark, and Carmike Cinemas) represent almost half of the total in the U.S. (National Association of Theater Owners, 2010). Many film theaters have operated in the hands of independent owners for a long time. However, independent theaters are increasingly being financed and distributed by the major studios and large exhibitors, and oligopolistic control never ceased to be a distinguishing feature of Hollywood (Aksoy & Robins, 1992). Many independent theaters have to work with major exhibitors in most countries.

Meanwhile, the transnationalization of theaters in non-Western countries has become peculiar, because major Western film chains also have substantial number of theaters in many other countries. For example, Cinemark has a sizeable number of screens in 12 countries in Latin America. This situation is not much different in other countries. In Korea, five major exhibition corporations, encompassing CJ CGV, Primus Cinema, Lotte Cinema, Megabox, and Cinus, owned 1,553 theater screens out of 1,996 (77.3%) nationwide as of December 2009 (Korean Film Council, 2010). Among these, Megabox, the third largest cinema chain, was sold to an Australian corporation in 2007 (Lee, 2007). Loew's Cineplex Entertainment also operates a division in Korea. As such, the concentration of ownership in exhibition has been noticeable in many countries, and several of them have been horizontally and vertically integrated by foreign exhibition corporations, in particular by Hollywood majors. While film producers and distributors in Western countries have increased their power in non-Western film markets, major theater chains in Western countries have directly increased their revenues through admission fees in their own or invested local theaters.

CONTINUING ASYMMETRICAL POWER RELATIONS IN THE GLOBAL FILM INDUSTRIES

Corporate integration in content industries has mainly taken the forms of vertical and horizontal agreements, which are powerful forces that reshape the media landscape. While vertical integration is traditionally a primary concern for film corporations, horizontal integration is also important in that it provides information of the concentration of ownership. These two different forms of integration cannot be separated, mainly because the production, distribution, and exhibition industries are closely linked in a complicated manner. Production indeed involves high levels of investment in a heterogeneous, highly perishable product, for which demand is uncertain, while exhibition involves the projection of that product to relatively small numbers of people in geographically scattered locales paying individually small sums that bear no necessary relationship to either the cost or the quality of the film. The film business has also been predominantly occupied by distribution (Garnham, 1990). Therefore, vertical integration, between different stages of the value chain and with content owners and distribution channels in a prominent position, and horizontal integration within the same level of industry work together to establish mega film corporations. The potential synergies created by these linkages across production, distribution, and exhibition, as well as across film corporations in the same category have made companies formidable players in information and entertainment (Collette & Litman, 1997). Major film corporations, in particular those in Hollywood, realized they could maximize their profits by controlling each stage of a film's life, and vertically integrated industrialization took the form of a studio system that in some ways made and distributed films in the same way that manufacturers make cars (Miller & Maxwell, 2006, pp. 36–37).

The structural transformations in the film market since the 1980s through corporate convergence indicated that competition quickly became the norm. Considering the economic and cultural impact of films on the public, the U.S. government and Hollywood have driven neoliberal reforms in many countries. Due to a series of deregulating markets, transnational capitals were active in domestic film markets around the world. In particular, all of the major film and television corporations in the U.S., including News Corporation, Time Warner, Disney, Viacom, GE, and Sony Pictures, have planned to invest in two emerging markets—China and India—in recent years. China, which had rejected foreign investment and foreign ownership in the communication industry until the late 1990s, permitted foreign companies to own up to 49% of Chinese video and audio distribution companies, as part of China's WTO accession in 2000 (Hazelton, 2000, p. 8). With the lifting of the ban on the global communication industry, several foreign majors invested in the Chinese communication market. The changing political-economic environment in the global cultural market has expedited the reform movement in the Chinese film industries.

In 2009, the Motion Picture Association of America (MPAA) also opened an office in Mumbai, India, under the name of the Motion Picture Distribution Association because the attraction of the Indian market is obvious, with growing numbers of movie goers and pay TV viewership. Dan Glickman, the chairman and CEO of MPAA, stated that the American studio would invest millions of dollars in the Indian film and television industry

(Indiaserver, 2009). The actual investment of American film corporations in terms of horizontal integration has not been phenomenal in these emerging markets; yet several U.S. film production corporations have acquired 12 Chinese production and distribution firms and 27 Indian film production and distribution corporations in recent years.

In addition to the capital market, the dominant position of U.S.-based corporations has intensified Hollywood's influence in the global film market. Thanks to horizontally and vertically integrated film conglomerates, decisions about film content have become more concentrated and rest in the hands of relatively few Hollywood majors. For example, most Asian and Latin American countries face competition from Hollywood for audiences. In Asia, foreign films alone accounted for 90% of Taiwan's box office revenue in 2004. About three quarters of Thailand's box office receipts went to Hollywood majors (Klein, 2003). In 2004, foreign films garnered 63% of the revenue of the Japanese domestic market, which is Hollywood's biggest foreign film market (Chung, 2007). China is currently not dominated by Hollywood; however, of the recorded total, the top foreign films entering the country, mainly Hollywood blockbusters in 2007, earned 45.1% of the gross box office (McDonnell & Silver, 2009). Hollywood has continued to dominate the global box office, taking more than 60% share of the international film market over the last decade, and Hollywood has increased its presence in several countries in the midst of neoliberal globalization (MPAA, 2009a; Pfanner, 2009). Hollywood's dominance in the global box office topped $29.3 billion in 2009 (PricewaterhouseCoopers, 2009).

In fact, the cultural industry has long been important to the U.S. government and Hollywood majors, but it has gained even more importance in recent years because the film sector is one of the most profitable industries for the U.S. economy in global trade (Miller et al., 2005). Given that much of the enormous revenues generated by the U.S. cultural industry have come from foreign markets, the liberalization of the global cultural market is very significant for the U.S. government (Magder, 2004, p. 385). The U.S. government and Hollywood claim that non-Western countries have liberalized and deregulated cultural sectors. They believe the cultural market should be left in the hands of free-market forces. Several countries have had to open their capital and cultural markets, resulting in the rapid increase of Hollywood's influence globally. The U.S. State Department has extensively supported Hollywood by driving other countries to open their cultural markets, which means the U.S. government has been deeply involved in the cultural trade issue by demanding that other governments should take a hands-off approach in the cultural area. The U.S. government has initiated and developed its cultural policy because capital investment and production in the cultural industries, such as film, music, and television programs, are among the most significant areas of the U.S. economy (Miller & Maxwell, 2006). The central theme of U.S. foreign cultural policy has been to expand a network of global trade based on its state power, which is among the strongest in the world. As the core of a liberalized trade regime, the U.S. can press its capital advantages to maximum effect (Jihong & Kraus, 2002, p. 423).

While the growth of Hollywood's dominance in non-Western countries' box offices has intensified, film corporations in many developing countries have not been able to increase their influence in the global M&A and film markets. Of course, if local film corporations in different stages are adequately explored, and the government's policy is wise, Western capital is not able to fully function as a conqueror, but it can be made use of to create a new condition in which the domestic film industry may even benefit (Su, 2010, p. 54). However, film corporations in non-Western countries have not taken a major role in practice because the U.S. has wielded its dominant capital and cultural power in the area of film. The U.S. has greatly influenced the changing map of the global film sector as the major player of neoliberal globalization. Both the capital market and the content market are mainly, perhaps only, for the U.S.

As such, film corporations in the U.S., as major capitalists, have dominated not only the capital market but also the cultural market (Coe & Johns, 2004). Through horizontal and vertical integrations, Hollywood has obtained a greater share of the market, and it has more bargaining power with suppliers, while gaining more control over the prices it can charge for its products. This situation has provided Hollywood with the fundamental reasons to maintain and continue its dominant position in the global film market. It implies that the alliance between neoliberal globalization and U.S.-based TNCs as capitalists has been a new trend in transnational political economy. Consequently, cultural products like Hollywood films help colonize a global audience and help form a hegemonic culture, which has threatened the existence of other cultures and the creation of alternative ways of life (Su, 2010). Whilst neoliberal cultural policies and media capitals converge, transnational film corporations have taken key roles and have penetrated the global film market with their capital and cultural products.

CONCLUSION

The global film industries have substantially changed since the mid-1980s, and they have grown through capital flow as well as cultural flow. The neoliberal cultural polices and Western capitalists have caused the rapid transformation of global film industries, and the global film industries are embedded in, and transformed by, a more complex web of multilevel network connections. The introduction of neoliberal economic policy, adopting the deregulation, liberalization, and privatization of communication systems beginning in the mid-1980s, followed by the WTO agreement of 1997, has resulted in the relaxation of foreign ownership restraints in the cultural industries and has expedited the swift transnationalization of the film industries through horizontal and vertical integrations in many countries.

Within the context of changing neoliberal cultural policies, major film capitalists have played a significant role in the film market. With deregulation in each government, Western-based TNCs have invested an enormous amount of money in the film industries in both developed and developing countries because they became highly profitable sectors of the world economy. Mega film corporations have extended their influence in the global film market through M&As and ultimately acquired a larger share of and larger profit from the global film market. The emergence of mega film companies through integration has been driven to allow big companies to control content and hardware together in order to enable them to maximize their value and profit. Although size and scale economies and industry structure are not the only key forces, they certainly play a major role in developing major film companies' global dominance, and the obvious frontrunner is Hollywood, as always.

In the film sector, three different industries have always been concentrated in Hollywood, or at best take the shift for granted (Bakker, 2005). Although the leading role of the U.S. has been slightly reduced in the capital market in terms of its proportion of cross-border deals in the midst of consecutive economic recessions and counter-hegemonic movements in several countries in recent years, it has not brought about an increasing role for film industries in developing countries. Neoliberal globalization does not solely mean the dominance of the Western countries (Keane, 2006); however, the key players of neoliberal globalization are still mega corporations in Western countries, including mainly the U.S. Although some developing countries, including China, India, Mexico, Chile, and Korea, have increased their roles in the cross-border deal market, inequality and imbalance in the film sector between Western countries and developing countries exists, as in other media industries. Hollywood is still the strongest force in the film industry with the infrastructure to distribute a major $200 million production around the world. And that is unlikely to change any time soon (Knight, 2010). Furthermore, the gap between a few Western countries and the developing countries remains quite large.

Part II

De-convergence of the Global Information Systems and Culture

6 Restructuring of the Global Telecommunications System

Over the last three decades, the telecommunications industry has experienced significant growth and change. The number of telecommunications companies has soared, and the market size has enormously increased in response to such factors as deregulation and liberalization in the telecommunications sector. Technological developments, in particular with mobile telephones, have greatly influenced the structural shift in the telecommunications industry. The desire of a large group of multinational customers to obtain fully integrated, end-to-end global telecommunications services from a single source has created the impetus for telecom firms to offer multiservice broadband and seamless worldwide telecommunications networks (Goldman et al., 2003). Telecommunications companies especially adopt cooperative approaches to network building, such as mergers and acquisitions, joint ventures, legal partnerships, and strategic alliances (Jin, 2005). Corporate convergence both domestically and globally has been a norm in the telecommunications industry because telecommunications corporations seek synergy effects. With the growth of the Internet, the role of telecommunications corporations has been crucial, and this has also resulted in the rapid growth of telecommunications convergence.

However, telecommunications companies have begun exhibiting symptoms of what appears to be the similar life-threatening disease starting in the early 21st century (Schiller, 2003). In many countries, including the U.S., the U.K., Canada, and France, hundreds of telecommunications firms have gone bankrupt over the last 10 years, and the situation is no better in non-Western countries. Telecommunications corporations in these countries have experienced severe financial difficulties with two consecutive financial crises in the early 21st century. From Western countries to developing countries, overcapacity and severe competition among telecom companies have brought endless bankruptcies and financial deficits to the telecom industry. Rapid growth contributed to industry consolidation during the last two decades; however, the recent downturn in the telecommunications industry has demanded new business strategies for telecommunications corporations (Jin, 2005). This means that several telecommunications corporations have recently pursued de-convergence strategies, such as spin-off and/or split-off strategies as well as counter-deregulation, to survive in the midst of the failure of telecom convergence.

This chapter examines the structural transformation of the global telecommunications industry. It explores the role of global telecommunications corporations by investigating the ways in which they have changed their strategies. It also analyzes the transnationalization of the telecommunications industry as a form of M&As, and why the global telecommunications industry has turned its attention to de-convergence strategies. Finally, it discusses whether newly developing de-convergence strategies in the telecommunications industry have become a solid paradigm replacing telecom convergence in order to determine future telecommunications policy directions in the 21st century.

TRANSFORMATION OF THE TELECOMMUNICATIONS INDUSTRY

Telecommunications has become a key to socioeconomic development either within a national, regional, or global context over the last two decades. Since telecom has functioned as the basic infrastructure for the emerging information technology (IT), as well as acting as the major driving force in economic development, construction of telecom systems is taking place on a large scale throughout the world (Jin, 2005). Due to its importance as national infrastructure, the telecommunications sector has grown exponentially in several categories, such as telephone lines, market size, and investment. In particular, the recent expansion of the telecommunications sector in the 21st century has been notable in wireless lines, while the expansion mainly occurred in landline telephone lines until the late 1990s. The number of main telephone line subscribers reached over 1 billion for the first time in history in 2001. Main telephone lines jumped from 142 million in 1960 and 272.7 million in 1970 to 689 million in 1995 and to 1.0 billion by the end of 2001 (United Nations Statistical Office, 1972; ITU, 2003b). Fixed telephone subscriptions increased for a while until 2006 when the number peaked at 1.261 billion subscribers, and it has decreased gradually to 1.159 billion subscribers in 2011 (ITU, 2012). In contrast to this, the increase in the number of mobile phones is far steeper than that of main telephone lines because the expansion began from a tiny base of the development. The number of mobile phone subscribers increased 12.6 times from 90.6 million in 1995 to 1.15 billion in 2002, and it exceeded that of the number of subscribers to fixed telephone lines (ITU, 2011, 2003a). Mobile telephone subscriptions have soared in the early 21st century, and, in 2011, the number of mobile subscribers reached 5.98 billion—5.16 times higher than the number of fixed telephone line subscribers (ITU, 2012) (Figure 6.1). The rapid growth of smartphones, including Apple's iPhone and Samsung's Galaxy, has been a major driver of the sudden growth of mobile subscribers in recent years. According to survey

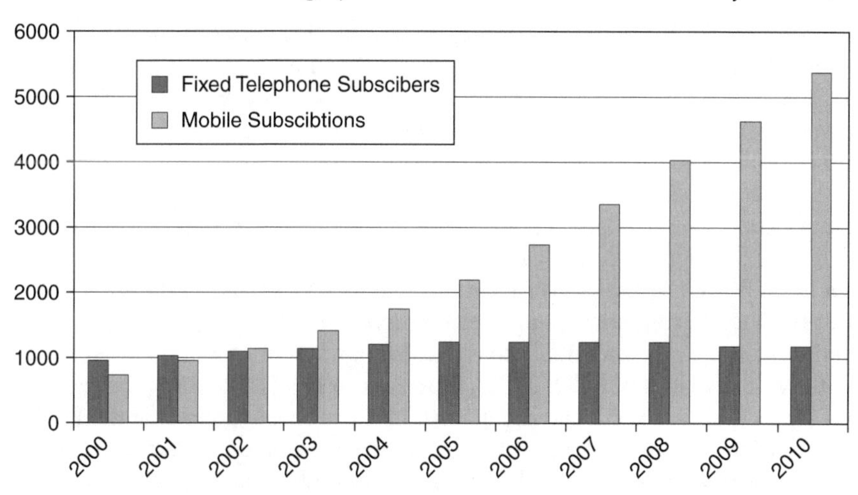

Figure 6.1 Growth of Telephone Lines, 2000–2011 (in millions)

Sources: International Telecommunication Union. (2011). *World Telecommunication/ICT Development Report 2011*. Geneva: ITU; International Telecommunication Union. (2012). *Key Global Telecom Indicators for the World Telecom Service Sector*. Geneva: ITU.

research (IHS iSuppli, 2012), in fact, smartphones officially crossed over the halfway point as they take over the mobile phone market in several countries, including the U.S., Korea, and Singapore, as of August 2012, and by the end of 2013, 54% of all phones globally will be smartphones.

Several Western countries, including the U.S., the U.K., France, and Japan, had dominated telecom businesses for several decades by the number of wired and wireless telephones until the late 1990s. However, the emerging market in recent years has also been remarkable in Asia, Africa, the Middle East, and Latin America, while Western countries' portion shrank, reflecting saturation of the market. In 1981, non-OECD countries (with 81% of the world's population in 1999) had only 7% of all phones (wired and mobile), but this increased to 11% in 1988 and to 43% in 1999 (Wellenius et al., 2000). This change especially resulted from the increase in the number of mobile phone subscribers, in particular through the use of smartphones in recent years, primarily in Korea, China, India, Brazil, Russia, and Indonesia.

Among these, China, due to the state's initiative for developing infrastructure for information technology, became the largest country in both the number of main telephone lines and cellular mobile phone subscribers in 2001 (with 144 million main telephone lines) and in 2002 (with 214 million cellular phone subscribers), followed by the U.S. (ITU, 2002, 2003a). In 2010, the number of China's mobile telephone subscribers reached 859 million, followed by India (752 million), the U.S. (278 million), Russia (237 million), Indonesia (220 million), and Brazil (202 million). In 2010 alone, the total number of mobile subscribers jumped only 2.5 times; however, it was as much as 214.8 times in India, 73 times in Indonesia, and 10 times in China, primarily because those countries started with almost no subscribers (ITU, 2011). Since the penetration rates in these countries are still much lower than in Western countries, the increase in the number of mobile customers will be soaring further in the near future.

As a reflection of the rapid growth of the telecom sector, global telecommunications revenues have been increasing. Global telecommunications revenues hit $1.85 trillion in 2010, a four percent growth from the previous year (Ng, 2011). Several Western countries enjoyed tremendous benefits from the growing telecom services market. For example, in 1999 revenues in telecom services markets in OECD countries accounted for as much as 89.8% of worldwide revenues (OECD, 2001). However, in the 21st century, the growth has been mainly driven by strong growth in the mobile segments of the BRIC nations (Brazil, Russia, India, and China) and a few emerging markets, such as Korea and Indonesia. The Middle East and Africa are to be the fastest-growing regions for telecom revenues over the next several years (Reed, 2008).

Phenomenal growth and transformation have become real partially because governments in several regions transformed telecommunications firms, from state-owned monopolies toward a competitive market structure. In other words, neoliberal telecommunications policy has expedited the transformation of the telecommunications industry. These forms of transformation are very significant because these strategies interact in determining the ownership of telecom firms and the industry structure of the whole sector, which results in concentrating the ownership structure to a few shareholders and board members, as well as conglomerating the telecom industries (Trillas, 2002). Telecom system build-outs, at every scale from local loops to the global grid, are occurring on an unparalleled scale throughout the world, although examples such as these are rare in developing countries (Schiller, 2001). In fact, during the 1980s and 1990s, maximizing shareholder value became the dominant ideology for corporate governance in the U.S. and elsewhere (Lazonick, 2009, p. 203). Since governments around the world were heavily responsible for the national infrastructure in the Fordism era until the late 1960s, the role of the government was prominent in many countries. However, since the early 1980s, governments have primarily pursued deregulatory measures in the public sectors, including telecommunications, and they have initiated liberalization and privatization of telecommunications corporations with counter-neoliberal activities in recent years.

TRANSNATIONALIZATION OF THE GLOBAL TELECOMMUNICATIONS INDUSTRY

The telecom sector is not just one among the network industries that experienced a policy paradigm shift over the last 20 years; however, it has been its core laboratory worldwide and the one where the reforms started earlier. If one had to pick a single year as the turning point, 1984 would be the most convenient one, with the parallel divestiture of AT&T in the U.S. and of British Telecom in the U.K. (Bacchiocchi et al., 2011, pp. 382–383). Since then, many states have driven neoliberal telecommunications policy, which has resulted in the structural change of the telecommunications industry. In fact, the telecommunications industry was operated almost entirely as a set of national monopolies, except in the U.S., prior to the mid-1980s, but each state has transformed the state monopoly system into a profit-driven private system over the last two decades (Cabanda & Afiff, 2002).

While there are several neoliberal measures in the telecommunications system, privatization and liberalization have been two key policy agendas for the transnationalization of the industry. To begin with, privatization has been significant because it involves the shift in legal status of an incumbent carrier from public to private ownership (Frieden, 2001, p. 86). Typically, but not always, privatization is also coupled with deregulation. The pace of privatization opportunities has increased, particularly because many states have made market access commitments, under the auspices of the WTO, which relax or eliminate foreign ownership and licensing restrictions in telecom. Privatization especially triggers the transnationalization of telecommunications corporations because governments need to open the domestic market to attract foreign investment as well. Korea Telecom—the Korean government owned telecommunications service industry, for example, was privatized in May 2002. The Korean government sold off its remaining 28.36% stake in Korea Telecom (KT) at that time, putting the final touches on the privatization of the telecommunications giant. The Korean government was able to complete the privatization of KT about 15 years after it first announced the plan in 1987 because privatization was not undertaken in a once-and-for-all manner, but at a gradual pace. The Korean government initiated the privatization, as well as the liberalization of the telecommunications industry, because of pressures from both national and international players in the late 1980s. There has been an increasing demand for participation in the Korean telecommunications industry. The pressure for the privatization of KT first came from foreign players, in particular the U.S. government, and thereafter, international organizations, such as the World Trade Organization (WTO) and the International Monetary Fund (IMF), as well as transnational corporations (Jin, 2005).

Meanwhile, liberalization has led the restructuring of the global telecommunications industry. Liberalization in the telecom sector can be traced back to the early 1980s when both the U.S. and the U.K. started their restructuring of the telecommunications industry. During the same period, AT&T in the U.S. had dominated the telecommunications market for several decades until the early 1980s, and the government and several telecommunications corporations tried to break up the situation through antitrust law. Going back to the early 1970s, in 1973, the U.S. federal government filed a lawsuit alleging that AT&T had: (1) illegally limited the kinds of connections and services MCI and others could get, and (2) illegally prevented other manufacturers from selling equipment to Bell companies. As a result, AT&T had to split off the monopoly telephone companies in 1984 (Weber, 2008). The breakup of AT&T ushered in a new era for the telecommunications industry: Deregulation and competition emerged as primary industry forces. This trend has continued into the 21st century in a variety of forms. The equipment market, which had been open to competition for several years, became truly competitive when the local telephone companies were no longer able to buy equipment from an affiliate company. The long-distance market also became fiercely competitive, paving the way for the deregulation of AT&T (Standard & Poor's, 2008, p. 19). In the early 1980s, the U.K. also privatized BT and allowed the authorization of Mercury, a competitive carrier, which was merged with three cable operators in the U.K. and named Cable & Wireless Communications in 1997. The changes that had been made in telecom market structures in the 1980s indicated that competition, once the exception in telecom, was quickly becoming the norm (OECD, 1990).

The 1990s indeed witnessed a tremendous transformation of the telecommunications industry. During the period, two major historical events, the Telecommunications Act of 1996 in the U.S. and the WTO telecom agreement in 1997, sent shock waves through the telecommunications sector and around the world (Schiller, 2003). In the U.S., the 1996 Telecommunications Act eliminated market entry barriers for entrepreneurs and other small businesses in the provision and ownership of telecom services and thereafter triggered a huge M&A trend across the greater communication industry. Since the passage of the Telecommunications Act, companies in the long-haul fiber-optic, cable TV, satellite, local telephone, wireless, and other sectors of the industry have undertaken massive capital expenditures to develop and upgrade networks (Bradbury & Kasler, 2000). At the global level, the 1997 WTO telecommunications agreement drove the liberalization process, which has resulted in the transnationalization of the global telecommunications industry. Signed by 69 countries, the Agreement on Basic Telecommunications Services requires those countries to open their domestic markets to foreign competition and to allow foreign companies to buy stakes in domestic operators (Jonquireres, 1997). The 1997 WTO telecommunications agreement especially exploded a wave of M&As between countries because several mega telecommunications corporations moved vehemently into deregulated telecommunications markets.

Consequently, telecom became one of the most active industries in the M&A market, which resulted in the formation of transnational mega corporations, beginning in the late 1990s. During the period 1982–2009, there were 33,291 mergers and acquisitions around the world, valued at $6,314.8 billion. As a reflection of the Telecommunications Act of 1996 in the U.S. and the WTO agreements in 1997, the majority of deals in the telecommunications industry occurred between 1996 and 2009 (89.4%). In particular, in the late 1990s, the world's largest telecom firms raced to put together global

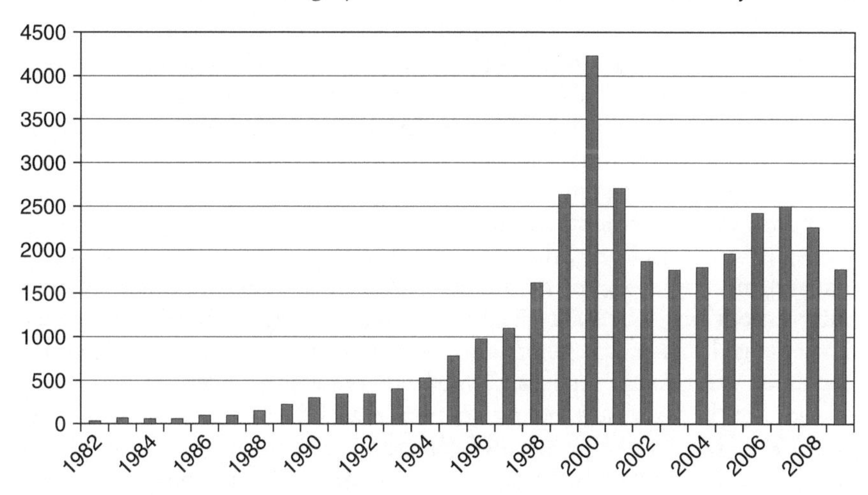

Figure 6.2 Global M&As in the Telecommunications Industry, 1982–2009

alliances. For instance, AT&T allied itself with Singapore Telecom and four major European national firms to form World Partners. Sprint, Deutsche Telekom, and France Telecom also formed Global One in 1997 (Maney, 1997). In fact, the deal market in the telecommunications sector peaked in 1999 (2,650 deals) and 2000 (4,238 deals).

The deal market, however, almost collapsed after the September 11 terrorist attacks against the U.S. and the following global recession in 2001. The number of deals fell to 1,768 in 2003, a 58% decline from the peak year of 2000. Although the telecommunications market had regained its strength in 2005 and 2006, it could not be the same as the peak season due to a consecutive financial crisis that happened in 2008–2009 in many countries. In fact, the number of deals in the telecommunications industry again fell to 1,789 cases in 2009. This implies that telecommunications corporations are very vulnerable to the financial crises (Figure 6.2). By country comparison, the U.S. has become the single largest country to purchase the majority of telecommunications corporations both domestically and globally. The U.S. has acquired as many as 13,103 companies (39.4%), followed by the U.K. (1,705 deals), Germany (1,678), Canada, Japan, France, and Australia. These Western countries account for 65% of the deals in the telecommunications market, which means that only a few countries have dominated the global deal market in the telecommunications sector.

The telecommunications industry has become significant with the rapid development of new technologies. Many media and telecommunications companies have sought the convergence of computers and telecommunications because this form of convergence is linked to, and partly responsible for, the convergence of once separated industries into a common arena providing electronic information and communication services (Mosco &

McKercher, 2006). New technologies also stimulate M&A activity. For example, the provision of broadband access services in the homes of consumers and the ability to combine content with transmission were key factors behind many telecommunications mergers. Further, the erosion of trade restrictions and other international regulatory barriers facilitated increased cross-border telecommunications activity, as well as a number of mergers and joint ventures among international firms (Goldman et al., 2003).

For example, in Korea, the telecommunications sector has been busy with convergence. Since the privatization of Korea Telecom—a public telecommunications firm—in 2002, the ownership of several major industry players has changed hands, and as of January 2011, only three major telecommunications companies remain and dominate the Korean telecommunications market. In the wired telecommunications market, KT accounts for as much as 84.3% of the market, while SK Broadband (13.2%) and LG UPlus (2.4%) are far behind. However, in the wireless market, SK Telecom is the leader at 50.5%, followed by KT (31.6%), and LG U Plus (17.8%) (Korea Communications Commission, 2012). In 2000, LG Telecom, LG Dacom, and LG Powercom were merged and changed their name to LG UPlus. KTF was merged with KT in 2009, which means that the Korean telecommunications market shows a very strong oligopoly system 10 years after the privatization of KT.

Since the Korean government turned its eyes to foreign capital when it privatized Korea Telecom in 2002, the government already revised the Telecommunications Business Act in September 2000 to increase the foreign ownership ceiling of KT from 33% to 49% (Jin, 2006; *The Korea Herald*, 2000). Foreigners owned 49% of the KT stock in January 2002 because KT was one of the most lucrative companies in Korea, and as of March 1, 2012, foreigners still own 49% of the shares, while foreigners have 42.5% in SKT and 16.8% in LG UPlus. Liberalization in conjunction with privatization has dramatically influenced the Korean telecommunications industry in the early 21st century. Korea's telecommunications industry has grown within the universal structure of networked global capitalism, and the state-led economy has been able to catch up to high-tech industrialization (Lee, 2012, pp. 14–15).

More interestingly, cross-border deals in the telecommunications industry comprised 9,318 cases, which accounted for 28% of all M&As. Cross-border deals were not significant until the mid-1990s. In fact, there were only four cross-border deals in 1983 and 51 deals in 1990. As in other media industries, the majority of cross-border deals have occurred since 1995. Cross-border deals completed among telecommunications corporations peaked at 1,376 cases in 2000, and about 90% of cross-border deals occurred during 1996 and 1997, right after 1996 Telecommunications Act and the 1997 WTO agreements. This data demonstrates that cross-border M&As in the telecommunications industry have rapidly increased with the relaxation of foreign ownership restraints and privatization since the mid-

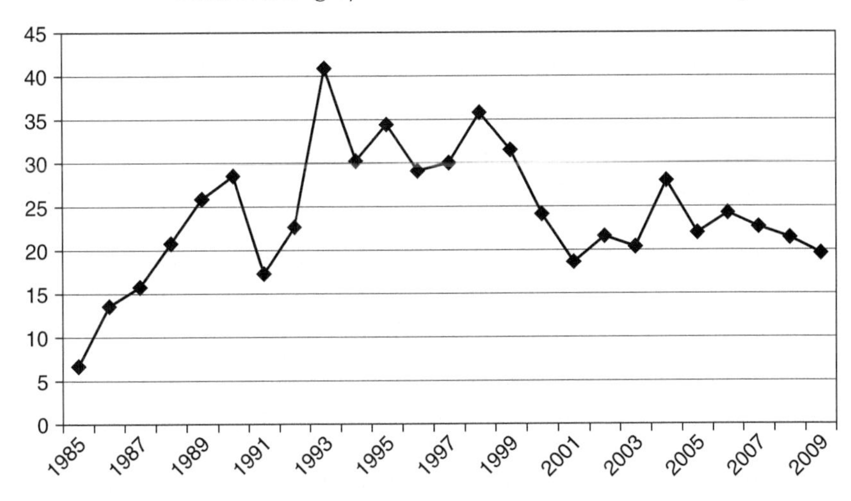

Figure 6.3 Cross-border Deals by U.S. Telecommunications Corporations

1990s. By country, the U.S. led cross-border deals with 2,318 deals, followed by the U.K. (768 cases), Germany (625), Canada, and France. These five major Western countries completed 4,627 cross-border deals, and this means that one of every two cross-border deals was carried out by these countries. Indeed, the neoliberal reform movements, first in the early 1980s and then in the late 1990s, which were driven by the U.S., have changed the map of the global telecommunications system via massive capital investment in the form of M&As. U.S. telecommunications companies have enjoyed supreme power in the midst of neoliberal reform by investing their capital not only within the U.S. but also in other countries (Jin, 2008).

However, the role of the U.S. in cross-border deals in the telecommunications industry has unexpectedly decreased over the last few years, particularly since the 1996 Telecommunications Act and the 1997 WTO agreements. The number of deals throughout the world has rapidly increased with these two major events; however, the role of the U.S. in the cross-border deal market dropped, with a few exceptions, right after these events. As Figure 6.3 shows, in 1993, the U.S. accounted for 40.8% of cross-border deals in telecommunications and 35% in 1998; however, this figure decreased to 18.6% in 2001 and to 19.6% in 2009. While the role of the U.S. has increased somewhat since 1996 in the telecommunications industry, the country has not sustained its capital power over the last few years. Although the U.S., as the major player in neoliberal globalization, has greatly influenced the changes in the map of the global telecommunications sector, the capital market does not exist solely for the U.S.

Of course, that does not necessarily suggest a significantly decreasing role for the U.S. because the U.S. overall accounted for 24.8% of cross-border deals in the telecommunications industry. A few Western countries still dominate the global telecommunications market with their capital, but the U.S. has slightly lost its supremacy in the global communication system.

FAILURE OF NEOLIBERAL TRANSFORMATION IN THE TELECOMMUNICATIONS SYSTEM

Since the early 21st century, several telecommunications giants, both service providers and hardware makers, have utilized de-convergence as a new business strategy. Instead of further pursuing convergence targeting synergy effects and economies of scale, these telecommunications corporations have strategically adopted de-convergence, which focuses on a few core businesses, primarily the Internet and cable areas, through deconsolidation. Global telecommunications giants headquartered in the U.S., the U.K., and Canada have sold parts of the company and split off and/or spun off their companies rather than integrating between the new and the old companies. While M&As among telecommunications companies have been the major trend in the midst of neoliberal reform over the last decade, spin-offs and split-offs are gaining momentum in the telecommunications market in the 21st century.

The shifting business strategy in the telecommunications industry has rapidly grown mainly because media convergence has caused several serious problems, such as plummeting stock prices and feeble content in the midst of economic recession in many countries but particularly in the U.S.; these are major hurdles for the survival and expansion of the communication companies. Facing a threatening crisis, several telecommunications conglomerates have pursued a de-convergence paradigm in order to restore their revenue structures in a different way from convergence and to hoist their company images through a new business strategy focusing on a few profitable and core businesses (Jin, 2005, 2011a).

In fact, many telecommunications companies have experienced unexpected problems, such as overcapacity and overcompetition, as well as mismanagement, including accounting scandals, in the midst of the transformation of the global telecom sector (Schiller, 2003). While telecommunications corporations were expanding, they took on enormous amounts of debt; one firm after another began having difficulty repaying these obligations and went into bankruptcy. Throughout the telecom industry, demands have failed to match expectations, businesses are losing money, stocks have plummeted, and a radical consolidation is in the offing (Starr, 2002).

Most of all, over-competition has caused serious consequences throughout the world. With deregulation and liberalization of the telecommunications sector, new companies entered the market at a fast pace. In the U.S., as of early 2001, approximately 700 companies offered long-distance services, although a select few dominated the market (Standard & Poor's, 2003). Many telecommunications firms went on a buying spree for Internet or telecom and cable assets as new strategic sectors because of the possibility to make huge profits. Their strategy hasn't worked and has eventually led to harsh consequences, including consecutive bankruptcies.

Structural transformation was supposed to create economic growth. Some of these firms have had some successes, but the industry was imploding. As Dan Schiller clearly points out (2003, p. 66), "long viewed as leading the way into the information age of productivity and enlightenment, telecom companies suddenly are presenting symptoms of what appears to be the same life-threatening disease." As Michael Powell, FCC chairman, explained in his July 30, 2002 testimony before the Senate Commerce Committee, many telecommunications corporations realized that there simply wasn't enough business to go around in the early 21st century, and they raced to gain market share in a burst of over-competition and price wars that drove down revenues. Powell clearly stated that "the hyper-competition and vicious price wars that precipitated today's burst stemmed from the exaggerated forecasts of demand from 'the Internet gold rush." Since Powell is a strong advocate of deregulation of the communication industry, his remarks in testimony would be considered significant during the telecom crisis in recent years (U.S. Senate Commerce, Science, and Transportation Committee, 2002, p. 16).

Overcapacity and overcompetition became the major problems haunting the telecommunications industry around the world in the early 21st century. In other words, overcapacity and overcompetition in telecom rendered the major telecom companies vulnerable (Schiller, 2003). From an economic viewpoint, convergence has not yet been a success, and this is for both coincidental reasons related to broader economic situation and structural reasons—management had great difficulty finding synergies between activities that are very different from one another. For example, exclusivity agreements between subsidiaries for the distribution of internally produced content could lead to new profitability, but could also prohibit external content (George, 2010, p. 559).

DE-CONVERGENCE AS A NEW PARADIGM IN THE TELECOMMUNICATIONS INDUSTRY

In the post-telecom crisis era of the early 21st century, many telecommunications corporations have pursued de-convergence strategies. In response to weaknesses in the long-distance market and depressed telecom stock prices, leading telecom companies began to separate their units, primarily wireless phone, Internet, broadband, and cable services, from more mature business operations in an effort to obtain higher market valuations and high profit (Jin, 2005). In other words, major restructuring projects among telecom carriers were related to announcements entailing increased decentralization, unlike ever growing concentration through M&A (OECD, 2001, pp. 14–15). What major telecom companies worried about was the lack of investor confidence and capital markets closing to new investment. The significant trend is for some of the largest telecom carriers to restructure themselves mainly due to commercial imperatives. In fact, up to the end of 2000, there was a flurry of deal making in the telecommunications industry. Successful transactions hit the headlines but many failed to materialize. Lack of institutional capacity, an absence of know-how, and political fallout all conspired to scupper the process. When deals were struck with domestic private-sector investors, the investors often struggled to raise finance (*The Banker*, 2003).

Regulatory constraints are an exemplary force leading to reorganization of businesses and creating opportunities for companies to float units. For example, in April 2000, the U.K.'s BT announced a restructuring of the company along its different lines of business (e.g., retail, Internet, wireless) rather than along geographical divisions (Sommerville, 2000, p. 7). BT indeed separated its service business from its network in January 2006 (O'Brien, 2007). In most European countries, fixed telephone services were completely opened for competition, thereby completing the liberalization of telecommunications services in most member states of the European Union (EU). However, contrary to what might be expected, these countries have increased their regulatory pressure on market actors rather than easing it, as a consequence of the consolidation of alternative operators and competition. In fact, the European Commission (EC) is proposing the creation of a new regulatory body, and regulatory pressure increases in the region (Herrera-Gonzalez & Martin, 2009).

In the U.S., Verizon Communications implemented a unique strategy, emphasizing the domestic market. As part of a strategic rebalancing to focus on growth opportunities in the U.S., Verizon completed the sale of its Latin American and Caribbean operations (in the Dominican Republic, Puerto Rico, and Venezuela) in March 2007 (Standard & Poor's, 2008). In addition, the recent economic crisis of 2008 precipitated the fall of the Canwest group, which has been one of the largest media moguls in Canada. The empire effectively collapsed in 2010 under the weight of a debt load of $4 billion due to two major purchases: newspaper group Hollinger in 2000 for $3.2 billion and the specialized television channels of Alliance Atlantis in 2007 for $2.3 billion, largely financed by the American business investment bank Goldman Sachs. Canwest subsequently found itself squeezed between the 2008 crisis and a decline in revenues, especially advertising revenue. Ultimately, the conglomerate was split into two groups: one for television and the other for print (George, 2010).

Above all, de-convergence as a form of spin-off/split-off has been common in many countries in recent years. For example, Alltel, America's largest rural telephone company, has spun off its landline business since 2006. This restructuring process allowed the fifth largest wireless carrier to focus on the fast-growing cellphone business and shed its less attractive fixed-line customers (Sorkin & Belson, 2005). Regardless of its effort to focus on a new business, however, it was sold to Verizon in 2007. In New Zealand, Telecom New Zealand's proposal to de-converge Chonus, the firm's fixed-line infrastructure arm, was approved by shareholders in 2011 (Global Telecoms Business, 2011). In November 2009, Cable & Wireless in the U.K. announced its intention to separate into two companies, Cable & Wireless Communications and Cable & Wireless Worldwide, reflecting the Board's belief that the two businesses had reached a position where they are best placed to deliver further value to shareholders as separately listed companies. The two companies separated in 2010 (Cable & Wireless Communications, 2010). Spin-offs in the telecommunications industry were pioneered in the U.S., but European corporations have embraced these strategies quickly. As will be detailed in Chapter 9, for example, AT&T has rapidly pursued the de-convergence strategy over the past decade. Of course, M&As have not ended because several telecommunications corporations continue to expand with convergence strategies. What has changed is that these companies aim to focus on wireless markets and the Internet.

It is certain that a common thread of the restructuring of telecommunications corporations has been the goal to build greater shareholder value. However, the mega mergers in the telecommunications industry in the 1990s and the early 21st century no longer impressed investors unless they targeted specific lines of business, in particular new media, including the Internet and wireless. As M&A entices investment in the new giant company, spin-off strategy also entices investments in telecom because new companies focus on profitable businesses, such as the Internet and broadband (OECD, 2001). In fact, investors apparently love seeing companies go to pieces. There's no mystery why companies make the financial maneuver. Investors have shown a clear appreciation for shares of major spin-offs. The 10 largest spin-offs between 2000 and 2005 had posted 41% stock market gains since being carved from their partners (Krantz, 2005).

Meanwhile, many countries still maintain managerial power over telecom industries as major stakeholders. Several developing countries, as well as developed countries, have chosen to keep some form of restrictions on foreign ownership because of the need for restricting foreign investment in the telecommunications sector. Overcompetition and a high level of debt especially required several governments to initiate counter-deregulation communication policy to encourage the formation of alliances to share facilities and infrastructure, resulting in a new mega telecom industry in the region (Standard & Poor's, 2002, pp. 11–12). There are a large number of state holdings in telecommunications operations, and privatization of telecommunications corporations still continues, although it is controversial in many countries. In China, the government has been concerned about its budding telecommunications industry that should compete with global telecommunications corporations; therefore, the government has utilized a relatively protective policy. It is very difficult for foreign investors to access China's telecom sector due to heavy regulation and policy restrictions on foreign investment.

After several failed attempts to sell Nigerian Telecommunications Limited (NITEL) in the midst of political uncertainty since 2001, the Nigerian government has canceled the sale of the state telecommunications company (BBC News, 2008). As one of the most recent actions, the Turkish government partially privatized Turk Telecom when Oger Telecom bought a 55% stake in 2005, and it planned to sell its remaining share through a public offering in the stock market (Global Telecoms Business, 2010). In the Philippines, the government has purposely encouraged the merger of regional telecommunications companies due to overcompetition. After acquiring two of the new players in the landline telephone sector, Philippine Long Distance Telephone (PLDT) has acquired control of two mobile operators, including Telesat Philippines and Smart Communications, making it the largest operator in the country. As of December 2011, PLDT owns four wireless and six wired line telecommunications corporations, either partially or wholly (PLDT, 2011). The consequences of neoliberal reform in many countries bring about new regulation to some governments in the market-oriented globalization era.

In sum, neoliberal transformation through liberalization and privatization has increased the number of companies and competition in the market for a while; however, both overcapacity and overcompetition, which were expedited by neoliberal globalization, became real culprits for the telecommunications crisis (Schiller, 2003). Liberalization and privatization of telecommunications industries in many countries were carried out too fast and too far, impairing the ability of telecom firms to establish viable businesses. Consequently, many telecommunications companies have started with new strategies to survive, including spin-off and/or split-off strategies as well as counter-deregulation in the early 21st century (Jin, 2005). The neoliberal transformation of the telecommunications industry is still occurring around the world; however, current forms of de-convergence, such as split-off and counter-deregulation within the global telecom sector, have gained power. Overcapacity and destructive competition have come to haunt the frontier of 21st century capitalism (Schiller, 2003, p. 66), and, due to the failure of neoliberal reform, a new transition period toward de-convergence has been brought about.

CONCLUSION

This chapter has analyzed the structural change of the global telecommunications industry through its implications of neoliberal globalization. The telecommunications industry has been a symbol of national infrastructure, and the global telecommunications industry has fundamentally shifted since the 1980s, and in particular since the mid-1990s. The changing political-economic environment owing to the two historical events (Telecommunications Act of 1996 and the 1997 WTO agreements) has rapidly changed the telecommunications industry primarily from government-dominated sectors to profit-driven private sectors. In the midst of deregulation and liberalization in many countries, many TNCs and financial banks worldwide have massively invested in non-Western countries because telecommunications were expected to become a highly profitable sector of global capitalism. In conjunction with big financial corporations who wanted to have telecommunications corporations due to the importance of telecommunications systems, major telecommunications corporations in Western countries have massively invested in non-Western countries. While many telecommunications corporations built networks, some of them burned cash.

In fact, several telecommunications corporations have benefited from corporate convergence; however, many telecommunications firms have not witnessed the growth of their corporations. These corporations admit that overcapacity and overcompetition after neoliberal reforms have brought about the telecom crisis, and they have pursued de-convergence strategies. The neoliberal reform has not been applied in many developing countries, and several governments have taken counter-deregulation measures. As governments around the world played pivotal roles in the transformation of the telecommunications industry, they, once again, have taken a pivotal role in setting a new telecom agenda in the midst of the failure of neoliberal reform in the telecommunications industry.

To conclude, the telecommunications industry has experienced a rollercoaster ride of change, from convergence to de-convergence paradigms. In the midst of market deregulation and liberalization in the telecommunications industry worldwide, the global telecommunications industry has vehemently pursued convergence strategies; however, with the failure of neoliberal regimes in many places, new strategies, such as split-off and counter-deregulation measures, have been necessary business models for survival. Of course, convergence in the telecommunications sector still remains powerful, and it will continue in the future. But the growth and development of the telecommunications industry cannot be fulfilled without de-convergence strategies as well. National governments in many countries, as well as telecommunications companies, understand the importance of de-convergence, and they will work together to set the agenda for telecom policies to survive and to grow in the midst of the changing political-economic environments of the 21st century.

7 De-convergence of the Internet and Software Industries

The Internet services and software industries (hereafter the Internet services industries) have disintegrated in the 21st century. They have been major targets in the global M&A market during the late 20th century and the early 21st century. The Internet services industries, referring to Internet-related companies providing Internet connections, services, and software to individuals and organizations which use these services for communication, however, have experienced a surge of de-convergence in recent years. The previous decade's impressive series of corporate expansions through M&As in the Internet services sector changed into an equally impressive series of corporate contractions, divestitures, restructurings, and bankruptcies. Several critical historical events may have caused the Internet markets to grow and incumbents to expand such technological advances along with regulatory reforms and a booming Internet market; however, the resultant size and diversity may actually have exceeded what integrated incumbents could efficiently manage under prevailing conditions.

In the early 21st century, incumbents have started to divest and withdraw, turning themselves into less diversified companies or even into entirely different organizations (Ulset, 2007). Some media and telecommunications corporations that have Internet services firms, including AOL-Time Warner, Vivendi, Bell Canada, and AT&T, have sought a new survival strategy, de-convergence, such as through spin-off and/or split-off strategies as well as sell-off in recent years. Instead of further pursuing integration and targeting synergy effects, these communication giants have adopted de-convergence, in order to focus on a few core business areas. As discussed in the previous chapters, M&As have been the norm in the midst of neoliberal reform, and the Internet services industries are attractive for many communication companies. However, as exemplified when Time Warner finally completed its AOL spin-off in December 2009, de-convergence is further gaining momentum as another significant trend in the Internet service industries as part of the communication market.

This chapter investigates how the transformation of the global Internet services and game software industries can be understood within the larger context of global political-economic shifts and accompanying neoliberal transformation. It examines what conditions set the agenda for media and telecommunications corporations to analyze a long-term transformation in the global Internet services system by analyzing consolidation through M&As in the Internet services sector. It also discusses the role of TNCs by studying how they are involved in the reshaping of the global Internet services and game software industries. It later calibrates why and how communication giants in Western countries, particularly the U.S. Internet services industries, have pursued de-convergence in recent years, while it discusses whether non-Western countries have increased their capital power in the midst of the shifting media ecology of the 21st century.

NEOLIBERAL RESTRUCTURING OF THE INTERNET SERVICES INDUSTRIES

With the rapid growth of technologies, in particular information and communication technologies (ICTs), contemporary society has witnessed a dramatic change. While traditional material-based industries, such as auto, furniture, and steel industries are still significant in national and global economy, ICTs have rapidly become a new area that corporations desire to own and control. Due to the significance of ICTs, many scholars have introduced several new concepts, and knowledge/information economy, information society, and network society have become buzzwords in our modern capitalist society.

As is well chronicled, Fritz Machlup (1962) introduced the notion of the knowledge industry, of which he distinguished five sectors: education, research and development, mass media, information technologies, and information services. Manuel Castells (2000, p. 21) emphasizes that one of the key features of informational society is the networking logic of its basic structure, which explains the use of the concept of network society. Whereas Castells links the concept of the network society to capitalist transformation, Van Dijk (2006, p. 20) points out that networks have become the nervous system of society, emphasizing the role of the network in the information society. Meanwhile, as Dan Schiller (1999b, p. xiv) points out, "networks are directly generalizing the social and cultural range of the capitalist economy as never before." Although there are some different focuses of these concepts, they emphasize the major role of ICTs in association with modern capitalism, toward which our society is heading toward.

Due to the significant role of ICTs, and in particular, the Internet, many communication firms have aggressively integrated the Internet services sector, primarily because the Internet has enabled the media and telecommunications industries to integrate with other industries. In the digital communication era, the Internet is a content distribution platform and a more effective means of transmitting voice, data, and video than proprietary networks; therefore, many communication companies have expanded their business areas through integration with Internet-related firms (Chambers & Howard, 2005; Liu and Chan-Olmsted, 2002; Li and Whalley, 2002; Chan-Olmsted, 1998).

While media and telecommunications have developed corporate integration and, later, disintegration, digital technologies and neoliberal communication policies have on a large scale facilitated the structural transformation of the Internet services industries. These two elements work in a complicated way, which has resulted in the change of ownership in the Internet services industries. Convergence between content providers, such as television and radio networks and transmission channels, such as those dealing with Internet service networks, has become the most important in the media and telecommunications industries (Jin, 2011a; Jenkins, 2006), and digital technologies and neoliberal communication policies have played major roles in the convergence process.

Most of all, major communication corporations worldwide have tried to obtain Internet-related firms through vertical and horizontal integration. Integration is especially challenging for many traditional forms of media that attempt to step into the new media sector, and pervasive digital technologies, including the Internet services industries, have fueled current media integration (Huang & Heider, 2008; Bar & Sandvig, 2008). As Mueller concisely points out (2004, p. 312), corporate integration means the digitization of all media forms and the adoption of compatible digital formats by all networks and information appliances. Communication corporations acknowledge that the path of integration could in the future lead to a world in which both voice and video become standard applications provided over a (broadband) Internet connection; therefore, having a stake in the Internet services sector is key for these companies (OECD, 2005, p. 210). The Internet has been understood as a new enabler for generating additional revenues and profits, chiefly due to its synergistic benefit of leveraging the assets of traditional media companies on the Internet (Kolo & Vogt, 2004, p. 23).

However, integration has brought not only economic benefits, but also negative consequences to corporations in many cases, because convergence has failed in producing promised synergy effects. As Castells pointed out (2001, pp. 188–190), "throughout the 1990s, media tycoons pursued the dream of integration between computers, the Internet, and the media. Yet, the business experiments on media integration carried on since the early 1990s had ended in failure. Most of the forms of integration did not make money. Indeed, traditional media companies are not generating any profits from their Internet ventures." Peltier (2004, p. 271) also argues that "it became obvious that the attempt to build media giants through M&As undoubtedly failed, because many media firms after M&As did not improve economic performance measured by profit margins." Meanwhile, based on their case studies of several media firms that have conducted divestitures, Alexander and Owers (2009, p. 105) concluded that many large media firms have had to address dysfunctional organizational structures resulting from previous acquisitions, and the best strategy is to divest the units that do not fit well with other parts of the firm.

De-convergence, again, mainly refers to the business activities in which communication companies strategically decrease the magnitude of their operations in order to regain profits and public images and/or to survive in the market with several business strategies. As this book defines convergence as the consolidation of firms within the industry influenced by digitization and probusiness government policies, it subsequently designates de-convergence as the form of disintegration of corporations by either selling parts of the shares to other companies or splitting off and/or spinning off of their companies. What have been overlooked in these discussions on de-convergence are the Internet services industries. Although previous papers have analyzed the divestiture of media firms, there is no distinctive work focusing on either the integration or disintegration trends of the Internet services industries. As noted, the Internet services industries have been the most significant sectors in the communication industries due to their unique roles in bridging the old and the new media. Likewise, the Internet services sector is a key business area, whether communication corporations choose to keep or eliminate these entities in the due course of disintegration. In the midst of the consecutive economic recessions in the 21st century, communication firms have reluctantly adopted new survival strategies, and current forms of transition indicate a new corporate policy in the Internet services industries and, overall, in the global communication industries. Therefore, it is crucial to understand the nature of de-convergence in the Internet sector.

CORPORATE INTEGRATION IN THE INTERNET SERVICES INDUSTRIES UNTIL THE EARLY 21ST CENTURY

The Internet boom of the second half of the 1990s seemed to herald the arrival of a new economy with its promise that, after the stagnation of the early 1990s, innovation in ICTs would regenerate economic prosperity. The sharp economic downturn in 2000–2002 called into question the new economy's ability to deliver on this promise (Lazonick, 2009, p. 1). Nevertheless, the Internet services industries were relatively new and were the most dynamic in the communication sector, not only because of their roles as interactive communication tools but also because of their functions as the platforms of several different services, including as new media formats by which to watch television programs and films. The Internet services industries have rapidly increased with the growth of broadband services and mobile Internet services. The Internet services market, which has been measured by the total revenues paid by Internet users to Internet services companies, has been phenomenal. The world's Internet services market, including wire, wireless, and mobile Internet services, was recorded at $214.6 billion in 2008, and it was projected to be $333.6 billion in 2013. The Asia-Pacific region is the largest market, and the American Internet market soared from $24.3 billion in 2004 to $40.4 billion in 2008 (a 66.7% increase) (PricewaterhouseCoopers, 2009).

Among these, broadband services consisted of the largest sector; they recorded $132 billion in 2008 and were projected to grow 9.3% annually until 2013. The modem Internet market has decreased, and it will decrease further. The modem market was worth $30.5 billion in 2008. As a new trend, the mobile Internet services market has grown, and three Asian countries—China, Japan, and Korea—made up 67% of the global market ($52 billion) in 2008 (PricewaterhouseCoopers, 2009). The market share of broadband services has rapidly increased, mainly as it provides diverse services connecting television and telecommunications services. Broadband services are getting more important because people are able to download and enjoy cultural products, such as films, television programs, games, and music. With the changing behaviors of consumers who move around to find new entertainment tools, communication corporations have integrated Internet services companies.

Under this circumstance, the field of Internet services industries has witnessed an unprecedented number of M&As in the global capital market. During the period 1982–2009, overall 48,460 cases of M&As in Internet services and software companies, valued at $4,634 billion, were completed worldwide. As discussed in Chapter 2, the M&As of the Internet services industries are the largest in both the number of deals and the total amount of transaction values in the media and telecommunications sectors. The transactions of the Internet sector were not substantial until the mid-1990s, mainly because the Internet was not commercialized yet; therefore the transactions that occurred before the mid-1990s were mainly in the software sector, not directly related the Internet.

The Internet services industries are latecomers in the communication market. In fact, there were only 362 deals in the Internet sector, valued at $24 billion, in 1994, and, again, they were mainly software-related corporations. However, the integration of the Internet services industries has soared since the mid-1990s with the commercialization of the Internet. In particular, when the U.S. relaxed cross-ownership restraints between telecommunications and broadcasting with the 1996 Telecommunications Act, communication firms in the U.S. jumped into the deal market. The 1997 WTO agreement, which demands the liberalization and privatization of the global communication market and industry, has especially driven M&As in most countries, and cross-border deals in different countries rapidly increased for a couple of years. Momentous technological changes in computers and telecommunications toward digitization coincided with and enhanced the rise of neoliberalism. This has opened new opportunities for communication corporations by fostering economies of scope and scale (Kelsey, 2007). As a consequence, the number of deals was recorded at as many as 8,037 cases ($913 billion) in 2000 (Figure 7.1). With digitization and neoliberal globalization, communication firms have massively invested in the Internet sector, which has been the most significant industry in the late 20th and early 21st centuries.

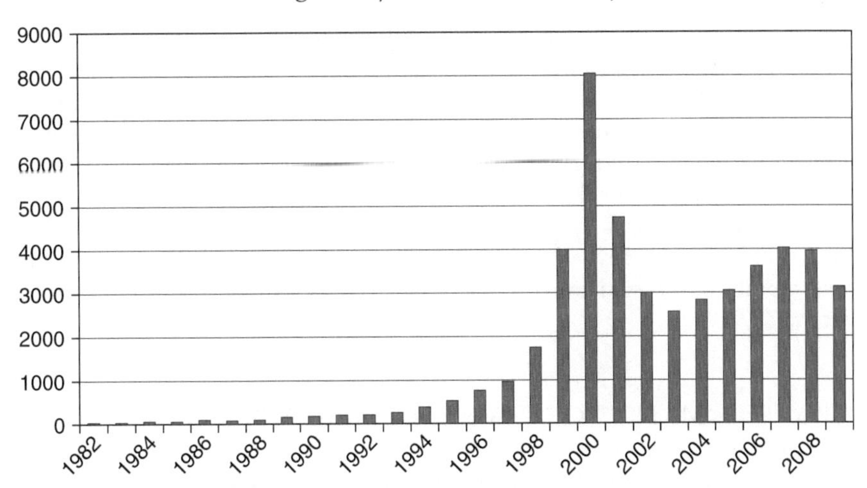

Figure 7.1 M&As in the Internet Services and Software Industry

The corporate merger trend in the Internet sector has not lasted long. While many communication companies still pursue investments in the Internet sector, M&As among the Internet services industries have substantially dropped off in the midst of the three consecutive economic crises that have occurred in the early 21st century. Most of all, the Internet services industries have been directly influenced by the dot-com bubble and the September 11 terrorist attacks against the U.S., both in 2001. These two historical events have resulted in the downfall of the U.S. economy, which has caused the global economic recession. The consequence has been severe. The number of deals in the Internet services industries dropped by 41% in 2001 from the previous year, while the total amount of transactions dropped 57% during the same period. This implies that the global economic recession during the period has both directly and indirectly expedited the decrease in the number of deals in the Internet sector. It is not easy to prove a direct relationship between the economic crisis and the decreasing number of deals; however, at least, it is certain that the economic crises have influenced communication firms because they have no capital to use for investing in the Internet sector in the midst of the crises.

In addition, some communication corporations, which had acquired Internet companies for synergy effects, did not secure what they wanted to get in the midst of the economic recession. What convergence brought was the decreasing value of their stocks, falling revenues, and devastating corporate images, which have resulted in changes of CEOs in many corporations, including AOL-Time Warner and AT&T. In fact, the number of deals in the Internet services industries has plunged, and there were only 2,548 cases in 2003 (see Figure 7.1). The Internet services industries had recuperated the trend of M&As for a while until 2007; however, it has dropped again, partially due to the economic crisis that occurred in 2007–2008 in the U.S. This trend proves that the Internet services industries have been more vulnerable than other media industries—especially compared to the content sector. For example, during the same period, the number of deals in the movie industries, including in production, distribution, and exhibition decreased immediately after the economic crises; however, they have returned to the same and/or even higher level of transactions. While the movie industry has not been much influenced by global economic downturns, the Internet services sector has been hit severely; therefore, disintegration in the Internet sector has rapidly burgeoned over the last 10 years, as will be discussed in detail later.

TRANSFORMATION OF GAME SOFTWARE INDUSTRIES

Other software companies, including game corporations, have shown a very interesting trend. During the period 1982–2009, 35,456 transactions in other software were recorded at $1,354.9 billion. While it is not clear how many companies are game-only software makers, since the majority of software companies are included in the Internet and software industries, it is not unsafe to say that the majority of them are game software makers in the 'other software' category. The transactions of other software were not substantial until the mid-1990s, as in the case of the Internet industry. There were only 28 deals in 1982 and 351 deals in 1990. As a reflection of dot-com boom in the late 1990s, the number of deals soared to 1,907 in 1999 and 2,963 in 2000 (Figure 7.2).

As in many media industries, including the Internet services industries, other software industries have been directly influenced by the dot-com bubble and the 2001 terrorist attacks against the U.S. The consequence has been severe. The number of deals in other software industries plummeted to 1,632 in 2002. This means that the global economic recession during the period has both directly and indirectly expedited the decrease in the number of deals in other software. Since the U.S. has been the largest acquirer in the global deal market in other software industries, it is certain that the economic recession in the U.S. has deeply influenced the global M&A trend. In fact, the U.S. alone consists of 45% of the deal in other software industries, which is the highest in the media and telecommunications industries. Following the U.S. are the U.K., Japan, Canada, Germany, and France, and these six major countries acquired 71% of other software companies. Due to the significance of Japanese console games, Japan acquired 2,029 corporations, the third highest. Korea, China, and India account for 6% of the global deals in this category, which is relatively higher than other industries, mainly because of their emphasis on the game industries.

Interestingly enough, the number of deals in other software has regained the rate observed at its peak point and has gone even further in the late 2000s. The number of deals was 3,414 in 2007 and 3,434 in 2008, again,

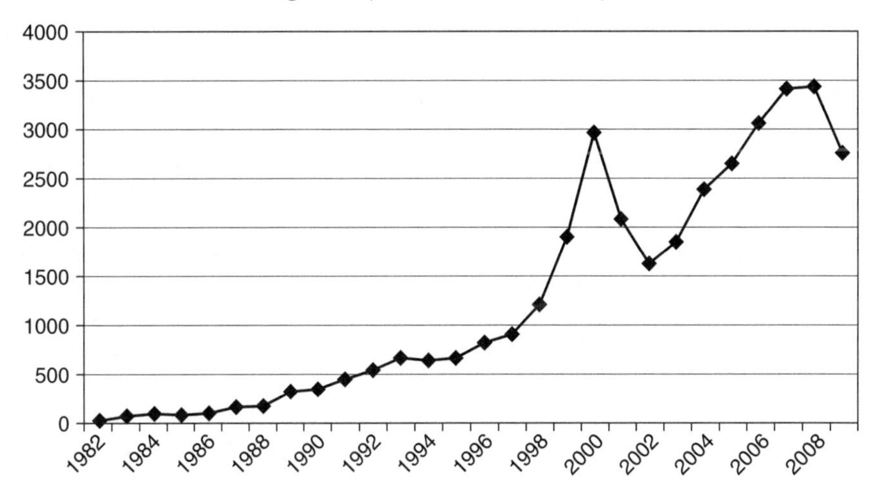

Figure 7.2 M&As of Game Corporations in Other Software

mainly due to the rapid growth of the game industries in the 21st century. This trend parallels the growth of the global game industries. Indeed, the video game market, as it relates to consumer spending on console, handheld, PC, mobile, and online games—referring to games carried out on a computer network—is a burgeoning new media industry (Jin, 2010; Dyer-Witheford & Sharman 2005). The video game industry has been one of the fastest growing segments; for instance, the growth rate was as much as 18.2% in 2008. Previously, the worldwide video game market was $27.8 billion as recently as 2004; however, it soared to $51.4 billion in 2008. It is expected that the video game market will increase to $73.1 billion in 2013 (PricewaterhouseCoopers, 2009, p. 44).

The primary driver of the global game market is the growth of console games, where gamers play on a dedicated console, such as Sony's PlayStation 3, Microsoft's Xbox 360, and Nintendo's Wii (Jin, 2010; Pricewaterhouse-Coopers, 2009). In the U.S., console games, including handheld games, constituted 74.4% of the market at $8.64 billion in 2007. The situation is not much different in Europe (60.6%), although online games are growing (PricewaterhouseCoopers, 2009). In contrast to this trend in many Western countries, online games have been markedly in vogue and are playing a critical role in Korea's systematic transformation toward a digital economy and culture. The growth of online gaming has been phenomenal in the Korean context. In 2000, for example, sales in the domestic gaming industry were valued at $1 billion, and online gaming only accounted for 22% of the game market. However, the Korean game market has dramatically changed in the early 21st century, because online gaming has become the most significant game genre and popular culture for youth. In 2011, for example, the market value of gaming, including console/handheld, online, mobile, arcade, and

PC games, was around $7.02 billion, up from $457 billion in 2009. Online gaming accounted for as much as 88.9% ($6.23 billion), followed by mobile (6%), console/handheld (3.8%), arcade, and PC games in 2011 (Korea Creative Contents Agency, 2010; 2012). As in the movie industries, the game industries have three major areas—developing, publishing, and distribution sectors—and they vertically and horizontally integrate to actualize synergy effects. As the most recent major deal in the game industries, Nexon, a Tokyo-based maker of online games, bought a $685 million stake of NCSoft to become the Korean company's biggest shareholder in June 2012. Nexon sees a lot of opportunities for acquisitions and is especially interested in investing in companies that develop games for Facebook and mobile phone users (Lee, 2012).

China is also an emerging online game market; therefore, a few Asian countries, including Japan, are relatively strong in the video game industries. In the most recent phenomenon, though, the game industry has focused on mobile games with the rapid growth of smartphones, such as Apple's iPhone and Samsung's Galaxy, in recent years. Due to the relatively cutting-edge functions and better screen quality of these smartphones, of course, in tandem with mobility, mobile gaming has rapidly become a major game genre in many countries, and mobile game producers and developers are gaining momentum in the global game market.

POWER SHIFT IN THE INTERNET SERVICES INDUSTRIES

There are several significant dimensions to the rapid transformation of the Internet services industries, which merit particular emphasis in the global context: the emergence of cross-border deals; the shifting role of the U.S.-based media and telecommunications firms as acquirer corporations in accordance to several economic recessions in the early 21st century; and the emergence of a few non-Western based capital investors, including China, in the global M&A market.

To begin with, during the period 1982–2009, the majority of deals in the Internet service industries occurred within the same country, while 23% (11,126 cases) of transactions were cross-border deals, in which acquirer corporations and target corporations were in different countries. Cross-border deals are critical because they indicate dominant power in the global deal market. By country comparison, communication corporations in the U.S. acquired the largest number of Internet services companies from other countries. Between 1982 and 2009, the U.S. acquired 3,471 Internet services firms (31.2%); thus, America's dominant position in the capital market is phenomenal. One country controls almost one third of cross-border deals in the Internet services industries, which have been the most lucrative and active sectors for many communication corporations. The next largest country is the U.K. (10.5%), followed by Canada. Following are Germany and France, and deals by communication corporations in these five Western countries consisted of 60.3% of total cross-border deals of the Internet services industries. This proves that the cross-border deals in the Internet services industries have been controlled by a handful of Western countries, as in other media industries such as film, newspaper, advertising, and broadcasting.

The role of the U.S. in cross-border deals has been the largest, as noted above. Communication firms in the U.S. have rapidly increased their investments through cross-border deals following the 1997 WTO agreement. Indeed, the country's proportion in the cross-border deal market consisted of 49% in 1998. Only one country (of course, the U.S.) accounted for almost half of the cross-border deals. The primary role of the U.S. in the global capital market has been made possible partially because the U.S. government, communication corporations, and financial firms have executed neoliberal globalization. They forced other countries to open their domestic markets, such as content and telecommunications. Many countries worldwide had no choice but to open their markets in the name of globalization as forced by the U.S. probusiness neoliberal communication policies that have lifted the barriers of new investment opportunities and resulted in the concentration of ownership through media integration into the hands of a few media giants, mainly U.S.-owned or U.S.-based corporations (McChesney, 2008; Jin, 2005). Thus, media corporations, including those in the Internet services industries in the U.S., have become larger and presumably more powerful. Although the role of the U.S. has slightly declined in the midst of the economic recession in the early 21st century, overall, the U.S. remains the largest capital investor in the global market.

Meanwhile, China has gradually increased its role in the Internet services industries. Although it is not prevalent yet, China (including Hong Kong) acquired 455 foreign Internet service providers (4%) and is the sixth largest force in the global deal market. China acquired only three foreign Internet services firms until 1997; however, the country has substantially increased its investment in other countries since the late 1990s. In fact, China acquired 135 foreign Internet services firms in 2000 and has consistently acquired about 40 companies per year during the 21st century. While other Asian countries, including Japan (277 cases), India (155 cases), and Korea (77 cases), have been far behind China in the number of Internet services firms they have acquired from foreign countries, China has become a major part of the global capitalist market and has increased its role. China has not been significant in its capital investment in other media industries, including broadcasting and film industries; however, Chinese corporations have merged with and acquired foreign Internet firms. With the 1997 WTO agreement, many Western countries have targeted China due to the importance of the Chinese communication market; however, China has also become involved in the global M&A market in the Internet services industries. Due to the importance of the Internet sector in the digital age, China has especially been active in the global capital market and has become an emergent force. Of course, this does not imply a shift of traditional power relations between the West and the East. Because other non-Western countries in Asia, Latin American, and Africa do not have enough power to compete with Western countries, America's dominant power remains.

DE-CONVERGENCE OF THE INTERNET SERVICES INDUSTRIES IN THE 21ST CENTURY

The Internet services industries have experienced decreasing value in the 21st century. Many communication companies had pursued integration with Internet-related firms, primarily because they believed that the Internet was a necessary tool in the digital era. However, some of these communication firms did not fulfill synergy effects due to several reasons, including cultural conflicts between acquirer and target companies and economic crises. Many communication corporations have especially shown financial difficulties because of overcompetition and overinvestment in the Internet services industries. In other words, convergence in these industries has caused several serious problems, which are major hurdles for the survival and expansion of communication companies. Therefore, they have shifted their business strategies and have transformed their companies through disintegration. Unlike the conglomeration of communication firms, they have utilized disintegration in order to slim down the size of their companies. Although disintegration is not new, the trend has gained momentum in recent years.[1]

Convergence, which has focused on the maximization of the profits of media companies through M&As, again, has not achieved intended results in many cases, and mega media giants have pursued a de-convergence strategy because they can neither get synergy effects nor lift their companies' images. As Manuel Castells pointed out (2001, pp. 188–190),[2] "the business experiments on media convergence carried on since the early 1990s have ended in failure. Most of the forms of convergence did not make money. Indeed, traditional media companies are not generating any profits from their Internet ventures." De-convergence in the Internet service industries has appeared in several ways: massive layoffs, sales of profit-losing companies, and/or spin-off/split-off, and in most cases, of course, they are not mutually exclusive. De-convergence has expanded particularly in the content and entertainment sectors more than new media areas because traditional media corporations did not get what they wanted through their mergers with the new media industry (Jin, 2007).

Several major media corporations, such as AT&T, CBS-Viacom, AOL-Time Warner, and Vivendi, have rapidly changed their corporate polices, from integration to disintegration. Although some of them are still interested in vertical and horizontal integration, many of them are seeking deconsolidation. Some media firms have emphasized content-focused companies after experiencing some difficulties due to the convergence with Internet-related firms. The primary example of content-focused disintegration is AOL-Time Warner, which completed its AOL split-off in December 2009. AOL-Time Warner had been a prominent practitioner of corporate synergy. When AOL and Time Warner merged with the highest ever transaction cost of $161 billion in 2001, they appeared a perfect match, and market expectations were high, even though the economy in the very early 21st century was showing signs of contracting.

However, as will be detailed in Chapter 8, AOL-Time Warner experienced significantly falling stock prices, which has resulted in the disintegration of the company (Musgrove, 2009). The company has been separating itself into four smaller media corporations. Time Warner planned to sell Time Inc., publisher of magazines such as *Sports Illustrated*, *Time*, and *People* (Kamitschnig, 2006). Time Warner spun off Time Warner Cable, which provides cable TV and broadband Internet access, in 2008. As its most recent move, Time Warner spun off AOL as a separate company in 2009. Time Warner said that with the separation of AOL, it is now better positioned to focus more closely on driving improved performance for its content businesses in the most effective way (*Internet Business News*, 2009). By spinning off AOL—the Internet services firm—and Time Warner Cable—which provides broadband Internet access—Time Warner has reorganized into a content-focused media firm, which would become a new model for several communication corporations. While many media and telecommunications corporations have tried to own the Internet services industry, Time Warner has strategically decided to find its future direction through the split-off of its Internet sector. Meanwhile, AT&T has structurally changed its ownership and corporate governance, from a divestiture harbinger to a merger giant, and finally to being a victim of corporate integration. AT&T became one of the first major communication companies to adopt disintegration strategies, mainly forced by the government. In 1984, AT&T broke itself apart by spinning off seven separate regional telecommunications operating companies and an R&D organization jointly owned by those seven companies.

It is clear that a common thread of this restructuring was the objective of building greater shareholder value. One conclusion that might be drawn is that the mega mergers of the late 1990s and early 21st century supported by neoliberal reform no longer impress investors unless they have targeted specific lines of business (Jin, 2005; Kolo & Vogt, 2004). As M&A entices investment in the new giant company, the spin-off strategy attracts investments in communication because new companies focus on profitable businesses, such as the Internet and broadband. By all means, the factors driving restructuring in the largest firms inevitably affect other communication companies that are still trying to be all things to all customers (Jin, 2005). More important, the integration behemoth itself had finally become a victim of M&As. As discussed previously, in the early 21st century, AT&T had been struggling with sharp declines in revenue, as price wars and increasing competition had sapped its strength (Latour & Young, 2005).

The trend toward voluntary media disintegration is no longer just a trend, but a full-fledged epidemic (McClintock, 2005), as AOL-Time Warner and AT&T are not alone in turning their backs on the mantra that big is beautiful, and several communication companies have followed what AOL-Time Warner and AT&T have done. Throughout the communication industry, demand has failed to match expectation, businesses are losing money, stock values have plummeted, and a radical consolidation is in the offing (Starr, 2002). Communication companies sometimes believe that spinning off an Internet venture is the best way to free themselves from corporate bureaucracy, infighting, and risk aversion. Only by spinning it off from the parent, the argument goes, can the Internet business compete with nimble dot-com startups to snap up dominant online positions on Internet time (Chavez et al., 2000, p. 20). These businesses believe that they are able to secure more customers when they establish networks; however, overcompetition among Internet services providers could not guarantee profits even though they have networks.

The disintegration models undertaken by Viacom, Time Warner, AT&T, Vivendi, and BCE are the largest examples of a growing number of companies spinning off units that are not central to their businesses. The disintegration trend is on the verge of being in full swing in the communication sector. The reasons toward voluntary media disintegration may not be the same, but the result is. Facing a threatening crisis, several communication corporations, in particular those who have Internet-related firms, have sought disintegration, which has resulted in focusing on a few profitable and core businesses, either content focused or Internet focused. In the midst of neoliberal reform, these corporations believed that they could utilize synergy effects through media integration; however, the recent wave of disintegration proves the problematic nature of neoliberal communication policies, which have resulted in the burst of integration in the communication industries.

IMPACTS OF DE-CONVERGENCE IN THE INTERNET SERVICES INDUSTRIES

De-convergence in the Internet service industries has changed the structure of the global communication industry by altering the traditional business model (Jin, 2011a). Most of all, convergence has become more complex in the early 21st century because many media corporations have primarily sought horizontal integration to increase market power and share the high cost of digital technologies (Iosifidis, 2005). Unlike previous large-scale consolidations, the communication industry is mainly interested in horizontal integration, such as consolidation among content-focused companies and/or new media companies, in the middle of the growth of de-convergence. Since the communication industry has learned that big media giants through vertical integrations may not be profitable, they have shifted their strategies. Instead of further pursuing M&As between traditional media and new media, the media industry has started to seek a new form of consolidation.

In fact, Comcast's bids for Disney in 2004, for example, did not work out. Comcast had a successful track record of acquisitions, but that was solely on the cable side, not the content side. Comcast, the nation's biggest cable operator, would own the ABC broadcast network as well as the Disney film studio, ESPN, and other Disney assets, if a deal was reached, as in the case of AOL-Time Warner. However, Comcast's purchase of ABC turned out to be just a dream because Disney was already worried about the potential problems that the deal would bring through other unsuccessful mergers (*CNN/Money*, 2004).[3] Instead, as in the merger cases between Google and YouTube (2006) and between Disney and Pixar Animation (2005), media convergence has focused on horizontal integration between production corporations and particularly new media corporations in recent years. Comcast did finally complete its bid to own NBC Universal in 2011, however.

Another major implication of the de-convergence trend is that media corporations emphasize core businesses, particularly content areas, instead of pursuing synergy effects through the integration between old and new media sectors. While the driving forces behind media convergence are often similar, the factors leading some to slim down are not as commonly shared (Fine & Johnson, 2005). However, as in the case of Viacom and Time Warner, they clearly sought a strategy to regain their strengths in the content area. 'Content is king' would be the mantra of the companies in order to survive and make profits. This change is quite unexpected given that one of the major reasons for the convergence of the media industry, between the old and the new, is the migratory behavior of media audiences, as Henry Jenkins (2006) observed. Media firms are having a presence on multimedia platforms because the convergence trend has been governed by shifting audiences' media habits (Jin, 2011a). The majority of media customers are unconcerned about media types; it is only the convenience and quality of content that matters to them. Once the consumers are guaranteed, media companies can catch the customer in any media option at any time or space (Chintala, 2008). That is why media corporations have vehemently pursued media convergence.

However, the media corporations now acknowledge that the convergence trend cannot guarantee the qualities of content that customers seek. They have witnessed that audiences are switching loyalties whenever they find better content. Digital technology is changing, and consumers' media habits are again migratory. Therefore, the questions of how to develop content companies and how to advance contents become major concerns for many media corporations. In fact, media corporations have rapidly become part of a digital capitalism that emphasizes the role of information within contemporary society by acquiring or establishing the Internet-related new media sector; however, this trend shows that the primary hallmark of the contemporary political economy has been a buildup of systemic overcapacity which has plunged the market system into crisis, as the failures of the integration model prove (Chakravartty & Schiller, 2010). Of course, it does not mean that media corporations focusing on content hold all companies under the same roof. When the merged media companies are no longer lucrative, they cut out these companies. For example, as one of the latest trends, Walt Disney, which bought Miramax (the art-house studio) in 1993, plans to sell it primarily because it is not attractive anymore. Miramax was valued at about $2 billion in 2005; however, after two Oscar-winning box office hits in 2007 with *No Country for Old Men* and *There Will Be Blood*, there have been no further big successes. As of April 2010, Bob and Harvey Weinstein, the brothers who found Miramax in 1979, are reported to be frontrunners to take the studio off Walt Disney's hands with a bid thought to be about $600 million (Frean, 2010). While many media corporations have emphasized content sectors, Disney is considering whether the mega-sized company postmergers can continue producing synergy effects, and it has decided to reduce the magnitude when it learned that size is not profitable anymore.

Finally, one of the major shifts in the de-convergence era is the change of the major players in the media market. De-convergence has indeed shifted dominant players in the communication industry. Over the last 10 years, the communication market has witnessed the emergence of media giants that have played a major role in the market. Communication convergence fundamentally shifts both market and completion conditions (Wirtz, 2001). Time Warner, Viacom, and Disney have dominated the media market, both domestically and globally. However, the media market has again witnessed a changing market order with the de-convergence trend. According to the *Fortune 500* (2008), Time Warner had been the largest media corporation in the market in recent years; however, with the separation of Time Warner Cable and potentially part of AOL in the near future, Disney would be the leader in the media market within a few years. Viacom, the second largest media corporation until the early 21st century, already ranked fifth, while CBS became the fourth largest, right after News Corporation, in 2007. As a result of these changes, the U.S. communication industry will witness a new era, in which several media firms compete with each other, while two or three big corporations, including Disney and News Corporation, dominate the market. This structural change in the media industry does not mean that major media giants will give up their dominance in the communication market and then move toward an era of the diversity of ownership structures in the near future. Instead, it mainly proves that these media moguls have pursued the de-convergence strategy, not because of the protests by civic groups, nor government regulations emphasizing media democracy, but due to economic imperatives. Indeed, the basic philosophy of de-convergence cannot be differentiated from convergence. Although the recent trends toward de-convergence have shifted the ownership structure of the communication industry at least for a while, the fundamental change in the ownership structure in the sector is still far distant.

CONCLUSION

The information services industries have witnessed a unique shift, from a corporate integration frenzy to the disintegration wave. Corporate integration had been a cornucopia in the communication sector over the last two decades, and integration has become a dominant paradigm in the information services industries in the midst of neoliberal reform. Corporations believed that media integration would bring synergies in the form of the rapid growth of profits. In particular, communication corporations have pursued integration with Internet services firms due to the importance of the Internet as a vehicle for the convergence of different technologies, industries, and cultures, as well as its symbolizing the convergence of the old and the new media.

However, the results of mega communication M&As emphasizing digital technologies amid neoliberal globalization have been quite disappointing. The major goal of M&As is the wealth maximization of shareholders by seeking gains through synergy, economies of scale, and better financial and marketing advantages. However, all this glitter seems to have faded. The ambitious expectations of communication executives and investors have evaporated (Kolo & Vogt, 2003). M&As in the information service industries have been unsuccessful in many cases, as many companies involved in mega communication deals subsequently became bankrupt or deconsolidated.

In the 21st century, the draconian media paradigm has partially lost its power due in part to overcompetition and overcapacity, which neoliberal communication policies have created, and the communication sector has tried to find a new business model. M&A activities in the communication sector are now rather driven by disintegration. Communication companies, such as Time Warner, Vivendi, AT&T, and CBS-Viacom, are amongst the largest sellers of assets in an effort to reduce their levels of debts. While experiencing the gloomy aspects of M&As, many communication corporations have sought disintegration strategies, such as split-off and/or spin-off. Communication firms may divest to reverse transactions that have not worked out, and the best strategy is to divest the units that do not fit well with other parts of the firm (Alexander & Owers, 2009, p. 104).

Disintegration, as a rapidly emerging media business model and strong force, is changing the communication market and ownership structure, although these changes were not the results of antitrust laws or from concerns of ownership concentration. However, as long as the current trend of disintegration happens primarily due to economic imperatives, the new trend of disintegration has not enhanced what the policy measures try to implement, at least in terms of corporations' intentions. Both merger and disintegration have commonly contributed to the profit-making process of commercial media corporations, instead of being dedicated to the growth of the public sphere. In other words, it is crucial to understand that the structural change in the information services industries does not mean that major media giants will give up their dominance in the communication market and then move toward an era of the diversity of ownership structures in the near future.

This work was supported by the National Research Foundation of Korea Grant funded by the Korean Government (NRF-2010-332-B00648).

8 De-converging Convergence in the Global Communication Industries

De-convergence has radically reshaped the communication landscape over the past decade. Just as convergence had become a major norm in the 1990s, de-convergence has now appeared as an emerging trend and has changed business practice in the communication system. De-convergence has become popular because most of the forms of convergence do not make money (Castells, 2001), and the business experiments on media convergence carried on since the early 1990s have ended in failure in many cases. In the midst of large scale failures of M&As in the 21st century, mega communication corporations have begun to find new business models, and many media corporations have strategically turned their interests toward de-convergence. While convergence is still a powerful corporate policy, corporate losses in revenue and profit with convergence have driven media corporations to shift their business paradigms from convergence to de-convergence.

This chapter examines the ways in which many mega media deals have failed instead of achieving promised synergies by investigating mega M&As that occurred between 1998 and 2007. It discusses why and how communication companies have pursued de-convergence by employing split-off and spin-off strategies as their new business models. It does not attempt to analyze microeconomic reasons for the failures of individual M&As. Instead, it maps out the macro-level trend of changing corporate strategies within the communication industries based on whether they have achieved synergy effects, such as increasing profits, revenues, and stock prices, or not. This means that the primary focus of the chapter is the contextualization of the changing corporate business trend by historically analyzing the rise and fall of media convergence and, thereafter, media de-convergence. I eventually articulate whether de-convergence could become a solid new trend replacing convergence.

CONVERGENCE AS AN OLD PARADIGM IN THE COMMUNICATION MARKET

While the convergence trend in the communication industries is not new, a process of convergence between the formerly separate broadcasting and telecommunications sectors has rapidly occurred due to the process of market liberalization and the digital revolution.

One can understand the changing nature of convergence in the communication industries through the rise and fall of mega deals. In order to empirically examine the current trend of de-convergence, this chapter analyzes 'The Top 100 Deals of the Year' in *Mergers and Acquisitions*, which is the dealmakers' trade journal, between 1999 and 2008. I examine major communication M&As within the top 100 deals of each year and then analyze whether they have achieved synergies or whether they have failed. When communication firms sell off/split off/spin off any consolidated companies or become bankrupt, we consider them as a failure because they did not achieve synergy effects, which is the most significant reason for pursuing M&As. Between 1998 and 2007, there were 103 communication M&As within the top 100 deals of each year; however, since one has to follow the results of convergence, I select 83 M&As completed between 1998 and 2003 because we must analyze the impact of the 1996 Telecommunications Act and the 1997 WTO agreement on the transnationalization of the global communication industry in the 21st century.

According to *Mergers and Acquisitions*, between 1999 and 2008 (transaction occurred between 1998 and 2007), among the top 100 deals of the year, including commercial banks and oil companies, there were a total of 103 communication deals, valued at $1,326 billion (Figure 8.1). As a reflection of the rapid growth of M&As right after the two aforementioned historical events (the 1996 Telecommunications Act and the 1997 WTO agreements), the number of deals and deal prices soared in the late 1990s. In 1998, there were 14 communication deals among the top 100 deals of the year (worth $112.3 billion), and there were 20 cases, valued at $302 billion in 2000. The mega deals in the communication industries numbered 13, and the total transaction price was $258 billion in 2001; however, the total number of deals of communication corporations significantly decreased to 3 cases ($14.9 billion) in 2007. The convergence trend of mega communication deals is exactly the same as the overall M&A trend of the communication industries.

More specifically, in the late 1990s and early 21st century, there were several mega deals in the communication sector. In 1998, the transaction between WorldCom and MCI Communication ranked fourth among the top 100 deals of the year. In 1999, three telecommunications deals made the top 10, including the deal between SBC Communications and Ameritech Corporation (the second highest deal of the year, at $62.6 billion). Of course, the biggest deal in the communication industry was the transaction between AOL and Time Warner. With a transaction price of $164.7 billion, this deal ranked first among the top 100 deals in 2001. The deal between AOL and Time Warner has been particularly important in the media market because it is a symbol of the future of media corporations. Many communication firms followed what AOL-Time Warner did to enjoy potential synergies. As in the case of AOL-Time Warner, many communication conglomerates believe they can secure the outlet of their content, including television programs and

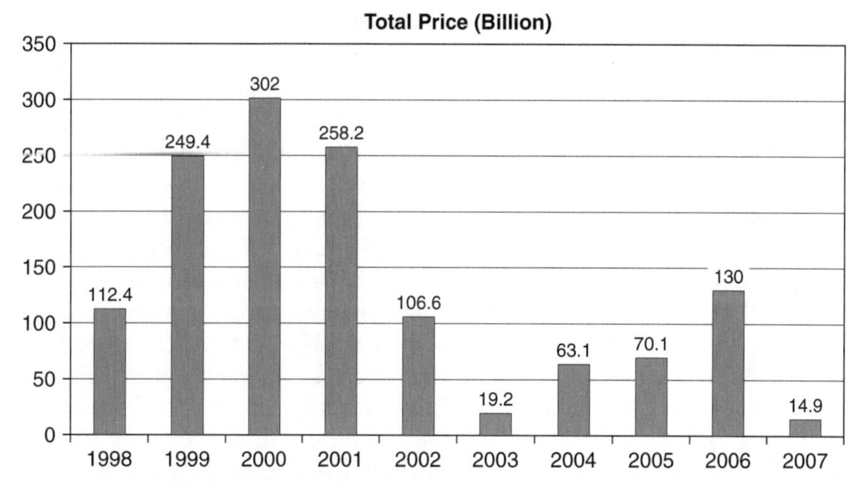

Figure 8.1 Mega Communication M&As Among the Top 100 Deals Each Year
Source: M&A Almanac, *Mergers and Acquisitions*, 1998–2007.

films, through vertical and horizontal integration, as well as international alliance (Jin, 2008, p. 370).

However, the numbers and values of mega communication deals have, all of a sudden, dropped, and this situation will not change in the near future, partially due to the current economic recession. In fact, there have been no significant mega deals in recent years other than the transaction between AT&T and BellSouth, which happened in 2006 ($72.7 billion). Media convergence as a form of M&A has gradually lost its grip. While we have been living in a burst of what passes under the name of convergence, we have started to witness the fall of M&As and soon thereafter the emergence of de-convergence, which will be discussed in detail later (Watkins, 2008).

A few political-economic reasons could be factors for the decrease in the number of national and global M&As in the communication industries: the terrorist attacks on the U.S. in September 2001; the economic recession starting in the early 21st century; and market saturation because too many deals occurred at once right after the Telecommunications Act of 1996 and the 1997 WTO agreements. Most of all, with liberalization and privatization, communication corporations have rapidly increased their investments in businesses; however, due to overcompetition and overcapacity, they have experienced the fallout of profits, which results in a significant decrease in convergence (Jin, 2005; Schiller, 2003). Communication companies rapidly jumped into the deal market through corporate convergence. However, with a few exceptions, many communication companies, including several of the largest conglomerates, such as Viacom-CBS, Vivendi, Time Warner, and AT&T, could not find the expected synergies, and they have begun to reconsider convergence—the golden rule dominating the communication industries over the last two decades.

FAILURES OF M&AS IN THE COMMUNICATION INDUSTRIES

It is not uncommon to hear that many M&As are not successful. According to data supplied by Thomson Financial, some 56,000 deals with a value of $6.4 trillion were announced between 1995 and 2000 (Adams, 2002); yet fewer than one half of those mergers survived. Historically, less than one half of mergers have survived, dating back to the inception of the merger process (Albarran & Gormly, 2004). The American Management Association examined 54 big mergers in the late 1980s and found that roughly one half of them led to failure in productivity or profits, or both (Chandra, 2001, citied in Mallikarjunappa & Nayak, 2007). In other words, the impact of M&As on stock prices, as well as on the operating profits in the period immediately following the mergers and also in the long run, have shown that M&As failed to create wealth for acquirer company shareholders, and they have often failed miserably (Mallikarjunappa & Nayak, 2007). However, there is no data available for the communication industries. Although the mass media report the failures of the mergers of communication corporations on a daily basis, only a few works of academic literature prove how many of the M&As in the communication sector have failed. Therefore, one cannot understand why and how communication corporations have changed their corporate strategies, from convergence to de-convergence.

To fill this gap, I analyze a total of 83 major communication M&As in the top 100 of each year (according to *Mergers and Acquisitions*) between 1998 and 2003. I define a merger failure in three different ways: (1) when communication corporations become bankrupt within a few years after the merger; (2) when communication firms split off and/or spin off acquired companies due to decreasing stock values and revenues; and (3) when acquirer companies are sold to other communication corporations due to lagging performances after the merger. It is not easy to confirm whether these failures are a direct result of M&As because there are other factors, including internal corruption (Adelphia Communications) and accounting scandal (Global Crossing). However, these factors are related to financial failure because they happened in the midst of M&As or right after the mergers, so these three major cases still represent the failure of the mergers.

Against this backdrop, 57 transactions out of 83 failed, which means about 68.7% of the mega transactions in the communication industries during this time did not achieve their desired ends, which are synergy effects (Table 8.1). This result shows that the incidence of merger failure in the communication sector is much higher than among all M&As mentioned in previous studies (about 50%, as in Adams, 2002), and it proves that M&As in the communication industries are much more vulnerable in the M&A market. New companies through M&As were expected to achieve several synergies that would deliver enhanced corporate images, economies of scale and scope, and increased revenues and stock values. However, as the results of the mergers explain, the majority of new companies after M&As have not achieved synergies, and their financial concerns increased.

Among failed M&As, 19 mega deals have bankrupted or closed down companies (22.9%), which means about one out of five mega deals experienced the worst kind of failure. From telecommunications corporations, such as WorldCom, Teleglobe, MCI, and Global Crossing, to broadcasting and newspaper firms, including Tribune Co.—the second largest newspaper publisher in the U.S., which acquired Times Miller in 2000—several businesses have filed for Chapter 11 bankruptcy protection. It is hard times for most newspaper companies due to losing daily circulation and declining advertising revenue. However, Tribune Co. had eight major daily newspapers and 23 television stations, so one can say that Tribune Co.'s aim to be a converged media giant, encompassing newspaper and broadcasting sectors, has failed (Ahrens, 2008). Yahoo!—an online giant—also acquired GeoCities (a web portal company) in 1999, Yahoo!'s second biggest acquisition behind Broadcast.com Inc.; it announced that it would close GeoCities in 2009 mainly because of Yahoo!'s financial difficulties (Vascellaro, 2009).

More specifically, in 1998, 14 mega deals occurred (including between WorldCom and MCI Communication, between Teleglobe and Excel Communications, and between Pearson PLC and Simon & Schuster). Among these, as many as 12 deals had failed within a few years after their mergers (Table 8.1). WorldCom, Northern Telecom (later Nortel in Canada), and Teleglobe (Canada) filed for bankruptcy protection in the U.S. and Canada in the early 21st century. In March 2009, Charter Communications, the U.S.'s third largest cable company—headed by Microsoft cofounder, Paul Allen (he acquired Charter Communications in 1998)—also filed for Chapter 11 bankruptcy protection. When it filed, the cable company said "it sought bankruptcy protection primarily because of the debt ($21.7 billion) it had accrued over years of expansions and acquisitions, not any operational issues" (de la Merced, 2009).

In 1999, 14 out of 20 communication deals among the top 100 of the year also experienced unwanted failures. Global Crossing went bankrupt, while GTE Corp., after the merger with Ameritech Corp., was sold to Bell Atlantic Corp. in 2000, which has since turned into Verizon Communication. The story goes on. In 2002, there were 13 mega deals, including the merger between Qwest Communication International and U.S. West ($56 billion), Bell Atlantic and GTE ($53.4 billion), and Viacom and CBS ($39.4 billion). Among these, 12 mergers failed, including the Viacom-CBS merger, the product of which was separated into two different companies starting in 2005. As is well chronicled, in September 1999, Viacom bought CBS, which was one of the largest network broadcasting companies. The merger brought together the extensive motion picture and television production, cable network, video retailing, television station, and publishing assets of Viacom, Inc. with the television network, radio station, and cable programming holdings of CBS, Inc., to create the world's second largest media

Table 8.1 Major Failed Media Corporations After M&As Occurred Between 1998 and 2003

Acquirer	Acquired	Transaction Price ($B)	Results of M&As
The Top 100 of 1998			
WorldCom Inc.	MCI Comm. Corp.	41.9	Bankrupt
AT&T	Teleport Communications Group Inc.	11.2	Acquired by SBC Communication, 2005
Northern Telecom Ltd.	Bay Networked Inc.	9.3	Bankrupt
Teleglobe Inc.	Excel Communications Inc.	6.4	Bankrupt
ALLTEL Corp.	360 Communications Co.	5.9	Acquired by Valor Communication, 2006
SBC Communications Inc.	Southern New England Telecommunications Corp.	5.8	
AirTouch Communications	MediaOne	5.7	Acquired by Vodafone, 1999
U.S. West Inc.	U.S. West Media Group	4.8	Acquired by Qwest, 2000
Pearson PLC	Simon & Schuster Inc.	4.6	Simon & Schuster was integrated with Paramount, 2002
Paul Allen	Charter Communications Inc. (90%)	4.5	Bankrupt
The Top 100 of 1999			
SBC Communications Inc.	Ameritech Corp.	62.6	
Vodafone Group PLC	AirTouch Communications	60.3	Acquired by Bell Atlantic, 2000
AT&T Corp.	Tele-Communications Inc.	53.6	Spin off
Global Crossing Ltd.	Frontire Corp.	10.1	Bankrupt
Clear Channel	Jacor Communications Inc.	6.6	
Adelphia Communications	Century Communications Corp.	5.2	Acquired by Time Warner and Comcast, 2006
AT&T Corp.	International Business Machines	5.0	Acquired by SBC Communication, 2005

Yahoo Inc.	Broadcast. Com Inc.	4.7	
Yahoo Inc.	GeoCities	4.7	GeoCities was closed
Chancellor Media Corp.	Capstar Broadcasting Corp.	4.3	
AOL Inc.	Netscape Communications Corp.	4.2	Failed merger; the brand name was removed
Cox Communications Inc.	TCA Cable TV Inc.	4.0	Acquired by Cequel III, 2006
Charter Communications Inc.	Falcon Communications LP	3.6	Bankrupt
BellSouth Corp.	Qwest Communications International (10.4%)	3.5	Acquired by AT&T Inc., 2006
The Top 100 of 2000			
Qwest Communication Inter.	U.S. West Inc.	56.3	
Bell Atlantic Corp.	GTE Group	53.4	
AT&T	MediaOne Group	49.3	Acquired by SBC Communication, 2005
Viacom	CBS	39.4	Separated into two independent companies
Clear Channel Communications	AMFM Inc.	23.1	
Bell Atlantic Corp.	Vodafone AirTouch PLC	15.0	
NTL Inc.	Cable & Wireless Communications PLC	11.0	Bankrupt
Tribune Co.	Times Miller Co.	9.2	Filed for Chapter 11 Bankruptcy, 2008
NTT	Verio Inc. (remaining 90%)	5.7	
France Telecom SA	NTL Inc. (25%)	5.5	NTL bankrupt, 2002
TeleCorp PCS	Tritel Inc.	4.9	Acquired by AT&T, 2002
VoiceStream Wireless Corp.	Omnipoint	4.8	Acquired by Deutsch Telekcom, 2001
France Telecom SA	Global One Co. (remaining 71%)	4.3	Global One was acquired by Eqest, 2001
The Top 100 of 2001			
America Online	Time Warner	164.7	Failed merger, spin off

(*Continued*)

Table 8.1 (Continued)

Acquirer	Acquired	Transaction Price ($B)	Results of M&As
Deutsch Telekom AG	VoiceStream Wireless	29.4	
Viacom	Infinity Broadcasting (remaining 35.7%)	13.6	Spin off
NTT	AT&T Wireless Group (16%)	9.8	AT&T Wireless was acquired by Cingular, 2004
News Corp.	Gemstar-TV Guide International (21.5%)2008	6.5	Gemstar was acquired by Macrovision,
VoiceStream Wireless	Powertel	6.2	Acquired by Deutsch Telekom, 2001
Exodus Communications	Global Crossing (GlobalCenter)	5.8	Global Crossing was bankrupt
Walt Disney	Fox Family Worldwide	5.2	
WorldCom	Intermedia Communications	4.2	Bankrupt
The Top 100 of 2002			
Comcast	AT&T Corp. (AT&T Broadband)	72.0	
Vivendi Universal	USA Networks	10.7	Acquired by GE, 2004–2005
AOL-Time Warner	Bertelsmann AG (remaining 49.5% of AOL Eurpow and AOL Australia)	6.3	Spin off
AT&T Corp.	TeleCorp PCS (remaining 77%)	4.8	Acquired by SBC Communications, 2005
Liberty Media Corp.	UnitedGlobal Com. Inc. (additional 65.5%)	2.8	
Bertelsmann	Zomba Group (remaining 80%)	2.7	Zomba was acquired by Sony, 2008
The Top 100 of 2003			
Liberty Media Corp.	QVC Inc. (additional 57%)	7.9	
AOL-Time Warner	AT&T Corp. (27.6% of Time Warner Entertainment)	3.6	Spin off

Source: The Top 100 Deals of Each Year, *Mergers and Acquisitions*, 1999–2004.

conglomerate (Waterman, 2000). However, Viacom started to separate CBS beginning in 2005, only a few years after the merger.

In 2001, the communication industries witnessed the largest of their transactions, between AOL and Time Warner. The same year, there were 13 mega deals; however, nine cases failed, including AOL-Time Warner, which has been separating itself into four different independent corporations. AOL and Time Warner appeared to be a perfect match, linking a major content provider with the distribution power of the world's leading Internet service provider. Because the two companies were established leaders in their respective markets, there seemed little doubt the merger would be successful (Albarran & Gormly, 2004).

As such, many mega communication corporations have been busy acquiring other media and telecommunications firms for synergies; however, the majority of communication corporations have failed to reach their goals. Neoliberal communication policies and the advancement of new technology have expedited the wave of convergence. The result is not so rosy for many communication firms. Communication corporations have reported failures one after the other, primarily caused by reckless mergers and acquisitions, and this situation will not improve in the midst of the contemporary economic recession.

Of course, the failure of M&As in the communication industries happens globally. Vivendi Universal Entertainment, a French conglomerate, acquired several communication companies, including Houghton Mifflin in 2001 and USA Networks and EchoStar Communications in 2002, but it was sold to GE in 2004, and it has become part of NBC Universal since 2005. To pare down its billions in debt, Vivendi Universal had to sell almost all of its entertainment divisions to NBC (Ahrens, 2004). BCE in Canada had also pursued media convergence to create a multimedia giant. However, as will be discussed in detail, BCE had to separate itself into several small companies.

Regardless of the failure of the majority of M&As, of course, there have been new transactions, including between Walt Disney and Pixar (2006), between News Corporation and Dow Jones & Co. (owner of the *Wall Street Journal*; 2007), and the merger between Google and YouTube (2006), which demonstrate that media convergence is still a powerful paradigm. We do not know whether these new deals are successful in utilizing synergies due to the short period of time after the merger. However, it is true that many major M&As have ended in failure, and the companies involved in them have now sought de-convergence as a new business strategy for survival in the market, which was unexpected. So the next question becomes whether convergence, as a form of M&A, will give way to de-convergence. Some investors already believe that mega mergers are out (Jubak, 2002). The skepticism following the failures of the mergers of several major corporations, including AOL-Time Warner, Vivendi, Viacom-CBS, AT&T, and BCE, keeps many investors and media professionals searching for a model that appears sound and, most important, profitable (Albarran & Gormly, 2004).

DE-CONVERGENCE: IS IT A NEW BUSINESS MODEL?

In the midst of the failure of mergers, several major communication companies have pursued de-convergence in order to regain profits, revenues, and shareholders' confidence. Convergence, which has focused on the maximization of the profits of communication companies through M&As, has not achieved the level of success it had been expected to. The de-convergence model, which is another form of corporate transformation, has appeared in several ways—sales of profit-losing companies and/or spin-off/split-off, but in many cases, these are not mutually exclusive. In fact, several major media corporations, such as Viacom-CBS, AT&T, and Vivendi, have rapidly changed their strategies, from convergence to de-convergence. Although some of them are still interested in vertical and horizontal integration, many of them are seeking de-convergence. This new wave for mega media corporations is very significant because their changing roles have immediately influenced the second-tier media corporations and small businesses to follow these mega corporations. In fact, the number of de-convergences has substantially increased in the early 21st century. Including spin-off and split-off together, there were 329 de-convergence cases in the broadcasting industry in 2002, but the number jumped to 524 cases in 2006, which is a 59% increase.

Different companies have diverse reasons for de-convergence. While the driving forces behind media convergence are often similar, the factors leading some to slim down are not as commonly shared (Fine & Johnson, 2005). For instance, differing cultures between two merged corporations, such as with AOL-Time Warner, have become a major reason for failure (Klein, 2003, as cited in Albarran & Gormly, 2004). Judgments about the inevitability of technological convergence were confounded by the fact that the two companies operated with entirely different corporate cultures. However, the investigation of these reasons is beyond the current analysis, so this chapter continues to focus on the documentation of the shifting trend from political-economic reasons, emphasizing synergy effects. To begin with, in 2005 Viacom-CBS separated the company into two independent media companies: one focusing on new media, the other on traditional media. This meant that a high-growth entity that retains the Viacom name was established (and it owns mainly cable and film sectors, including MTV, BET, and Paramount Pictures), while CBS owns the slower-growth parts of the business that includes television and radio broadcasting, as well as publishing and outdoor advertising (Sherman, 2006).

When Viacom broke up its company, the major reason was to maximize profits, as was the goal for the merger in the first place. In other words, Viacom considered economic aspects of de-convergence most, instead of diversity and media democracy (Waterman, 2000, p. 532). The breakup is a matter of pure economics: Share prices are languishing, and the media conglomerate is a concept that might have had its day. In fact, when the merger between Viacom-CBS occurred, Viacom shares were trading at $46.3 and peaked two years later at almost $75; however, they plummeted to $38.8 in March 2005 (Teather, 2005). Sumner Redstone, CEO of Viacom, broke up the media company on the belief that two separate companies could be worth more than the combined entity (Sherman, 2006). Redstone, who built Viacom, told the cable news channel CNBC that "synergy," if not dead as an idea, was certainly in its "death throes" (Teather, 2006, p. 22). While introducing the new business model, he stated:

> convergence, the bigger-is-better concept that dominated the industry for most of the past 10 years, is falling apart. As some of you know, divorce [de-convergence] is sometimes better than marriage [convergence]. (Maich, 2005, p. 33)

Although he did not explicitly say so, Redstone was marking the beginning of the de-convergence age—a new direction that repudiates much of what his audience was already grudgingly accepting (Maich, 2005).

The failure of AOL-Time Warner has especially sent shock to the communication market, because AOL-Time Warner has been a prominent practitioner of corporate synergy. As noted, AOL and Time Warner appeared a perfect match, and market expectations were high, even though the economy in the very early 21st century was showing signs of contracting. The notion was that the merger would unite distribution, provided by the Internet firm, with content provided by the media company. The justification was that ultimately all media distribution would take place over the Internet. More than this, the merger signaled the beginning of the era of full-scale integration, when personal computers would become TV sets, and TV sets would turn into computers (Anthony, 2009). The merger between Time Warner and AOL was indeed heralded as a surprising and revolutionary marriage between the old media and the new media.

AOL-Time Warner was expected to generate $40 billion in revenues and $11 billion in cash flow in its first year (Yang & Lowry, 2001); however, the new entity never met this expectation. Instead, AOL-Time Warner experienced significantly falling stock prices; the stock-market value of the media enterprise dropped by more than $100 billion between its merger and mid-2006, so the company has been separating into four smaller media corporations. Time Warner first sold its music company in 2004, and it quit the book industry in 2006. Time Warner (AOL was dropped from the company's name in 2003 in a symbolic admission that the mega merger might have been one of history's biggest corporate mistakes; *Times*, 2003) also plans to sell Time Inc., publisher of magazines such as *Sports Illustrated*, *Time*, and *People* (Kamitschnig, 2006).

Time Warner already spun off Time Warner Cable, which provides cable TV and broadband Internet access, in 2008. Time Warner stated that the media conglomerate decided to separate Time Warner Cable due to management and profit issues of the shareholders (Time Warner Cable, 2008). Jeffrey Bewkes, CEO of Time Warner, bluntly dismissed the notion of 'corporate synergy' embraced by his predecessors. Bewkes said that "cooperation between divisions should be encouraged, but no division should subsidize another, so the company is selling unprofitable businesses" (Kamitschnig, 2006). Time Warner decided to spin off AOL as a separate company by the end of 2009, and the separation of AOL is crucial because the spin-off emphasizes a shift from seeking size and scale—two attributes that were in vogue 10 years ago—to a focus on being nimble and innovative (Associated Press, 2009a). Bewkes believes that AOL's takeover of Time Warner was the ultimate failure in media convergence, and he plans to overhaul Time Warner as a profit-making media company (Wee, 2007).

In the telecommunications sector, two major cases, involving both AT&T and BCE, are worth investigating due to their major roles in North America. To begin with, AT&T, the largest telecommunications company in the U.S., has structurally changed its ownership and corporate governance. AT&T has been a convergence behemoth in the deal market. Between 1998 and 2007, AT&T had acquired or been acquired at least once each year. For instance, AT&T acquired Teleport Communications Group (1998), Tele-Communications and International Business Machines Corp (1999), and MediaOne Group (2000). However, AT&T has had to employ several different de-convergence strategies since the early 21st century as part of its restructuring process. AT&T had to overhaul the company because its total revenue dropped by 10.4%, from $42.1 billion in 2001 to $38.8 billion in 2002 (AT&T, 2002; Uchitelle, 2002). That was only the first signal of the de-convergence strategy in use by AT&T. Due to declines in long-distance voice revenue as a result of overcompetition in the midst of liberalization and privatization, AT&T had no choice but to divide the company into two sectors: old telephone services and new telecommunications services, including broadband (Jin, 2005).

In particular, when the company witnessed the rapid decline in investment toward AT&T after the merger, it started to massively pursue a de-convergence model in order to regain profits during this time of restructuring. AT&T believed that it would be able to take back corporate profit through the disjuncture of new media sectors, including mobile, Internet, and broadband, from domestic and international telephony businesses. In October 2001, AT&T therefore completed its split-off of Liberty Media Corporation as an independent company, which was a new media sector (in cable, including Discovery Channel). AT&T merged with Liberty Media in 1999; however, within only three years, it had to separate the company (Evers, 2000). The company also spun off AT&T Broadband to AT&T shareowners in November 2002. The convergence behemoth had finally become a victim of M&As. In 2005, the old AT&T Corp. was acquired by SBC Communications, although SBC has decided to turn its name into AT&T, Inc. (Belson, 2005). AT&T is not alone in turning its back on the mantra that big is beautiful, as several other telecommunications companies have followed what AT&T has done.

Meanwhile, BCE, a Canadian telecommunications giant, has shown a more dramatic structural change. When Jean Monty took over the job of CEO in 1998, he clearly pursed a convergence business model, attempting to combine both content creation and distribution within BCE, as well as taking greater advantage of the emerging Internet market. Jean Monty believed that in order to survive in a changing technological landscape, BCE had to have control over content. Shortly after the AOL-Time Warner merger, BCE acquired the CTV television network, the country's second largest network, behind CBC—a move to give it more programming to carry into battle against cable television and Internet services (Pritchard, 2000b). In 2001, BCE also acquired the *Globe and Mail*, the Toronto-based national newspaper, and combined it with CTV and the Sympatico-Lycos portal, its other content creation assets, to form Bell Globemedia. BCE's move reflects the broader scramble underway by telephone, Internet, and cable TV companies to offer content in packages of telecommunications services to consumers.

In the same year, BCE took a big step toward its goal of expanding globally with a $6.7 billion stock deal to buy the 77 % of Teleglobe that it does not already own and invest in Teleglobe's network linking 160 cities around the world (Pritchard, 2000a). As Dwayne Winseck argues:

> content, journalism, and all organizational resources strive to cybernetically integrate audiences into a self-referentially enclosed information system governed by the need to defend investments not just in networks and content, but a model of media evolution that has, at best, weak cultural foundations. The aim is to keep users within designated zones of cyberspace through the creation of content and service menus, the organization of hyperlinks, the bias of search engines, network architecture, synergies between content, elimination of alternative paths to somewhere else and so on. (2002, p. 811)

BCE quickly bought into convergence mania. However, the acquisition was a disaster, as BCE lost billions of dollars financing Teleglobe (CBC News, 2002).

As such, the result of mega convergence has not brought what this trend promised for BCE. Instead of increasing profits, BCE has experienced serious financial setbacks, and the value of its stocks has plummeted. Therefore, BCE made no secret of the fact that the company sees no reason for the phone and Internet company to own traditional media assets (Maich, 2005). BCE dramatically reduced its share of Bell Globemedia (now CTVglobemedia Inc.) from 68.5% to 20% by selling its shares to new shareholders in December, 2005 (Bell Canada Enterprise [BCE], 2005). In order to boost shareholders' fortunes, BCE has also reluctantly decided to dissolve itself by approving the $35 billion takeover of BCE by the private investment arm of the Ontario Teachers' Pension Plan and its U.S. partners, including Merrill Lynch Global Private Equity (Prashad, 2006). Although the privatization plan was blocked because BCE's auditor declared that the company would be insolvent if the transaction went through (Kouwe, 2008), it would become one of the largest takeovers in Canada (BCE, 2008).

The current CEO of BCE, George Cope, has set forth a more significant restructuring plan.[1] Mr. Cope acknowledges that Bell is a sprawling communication company with interests in telephone, satellite, internet services, and media assets (including the *Globe and Mail*), and, as the company has expanded outside of its core telephone business, it has come under criticism for the spotty quality of its telephone and Internet services. Therefore, a new CEO is considering selling its remaining interest in CTVglobemedia Inc. (Mcnish, 2008).

These de-convergence models are the largest examples of a growing number of companies spinning off units that are not central to their businesses. As of May 2008, large media companies have spun off 11 units that now trade on major U.S. stock exchanges, and in 2009, there were 30 spin-offs, the most in at least a decade (Krantz, 2008). The trend toward voluntary media deconsolidation is no longer just an exception but a full-fledged epidemic in the communication industries. The shifting business strategy in the communication industry has rapidly developed mainly because media convergence has caused several serious problems, such as plummeting stock prices and feeble content in the midst of economic recession in many countries, which are major hurdles for the survival and expansion of the communication industries. Facing a threatening crisis, several communication corporations have sought de-convergence, which has resulted in focusing on a few profitable and core businesses.

However, this structural change in the media industry does not mean that major media giants will give up their dominance in the communication market and then move toward an era of the diversity of ownership structures in the near future. Instead, it primarily proves that these media moguls have pursued the de-convergence strategy, not because of protests by civic groups, nor government regulations emphasizing media democracy, but due to economic imperatives. Indeed, the basic philosophy of de-convergence cannot be differentiated from convergence. Either convergence or de-convergence paradigms, "media corporations reproduce corporate ideology by presenting the public interest as synonymous with business interests and privileging consumer activity over citizen involvement" (Murdock, 2011, p. 81), because their primary concerns are increasing corporate power through the de-convergence paradigm, as they have done through media convergence. Although de-convergence has shifted the ownership structure of the media industry from concentration to de-concentration, at least for a while, the fundamental change in the ownership structure in the media sector is still far distant. Regardless of the breakup of the largest media firms as a form of de-convergence, the current trend does not indicate a shift of control from corporate suppliers to citizens.

CONCLUSION

The communication industries have shown a shifting trend, from a wave of convergence to one of de-convergence as a form of deconsolidation of the communication sector. Convergence has been strong in the communication sector over the last two decades, and it is well recognized that convergence has become a dominant paradigm in the communication industries in the midst of neoliberal reform. 'Big is beautiful' became a motto for media conglomerates because they believed that media convergence would bring synergies in the form of the rapid growth of profits.

However, the results of mega communication M&As are quite disappointing from an acquirers' shareholder perspective. The objective of M&As is, again, the wealth maximization of shareholders by seeking gains in terms of synergy, economies of scale, and better financial and marketing advantages. Media convergence is especially crucial for the old media because corporations believe they have to integrate with new media in order to please changing audience behaviors.

Despite high hopes, the majority of M&As did not achieve synergies, and many of the involved companies went bankrupt. An examination of the M&As of 83 mega deals completed between 1998 and 2003 proves that measured synergies were overstated. M&As in the communication industries, driven by digitalization and neoliberal reform, have been unsuccessful, as more than 68% of companies involved in mega communication deals subsequently went bankrupt or were deconsolidated. From media corporations, such as Vivendi, Viacom-CBS, and AOL-Time Warner, to telecommunications firms, including AT&T and BCE, the communication industries have witnessed the failure of mergers. As Bob Daly, chairman of the Warner Bros. studio, succinctly stated, "there was not a lot of synergy that took place" (Kamitschnig, 2006).

After experiencing the gloomy aspects of M&As, many communication corporations have pursued the de-convergence strategy, mainly through split-off and/or spin-off strategies. In the 21st century, the draconian media paradigm driven by both neoliberal communication policies and digital technologies has partially lost its power due in part to overcompetition and overcapacity, which neoliberal communication policies have created, and the communication sector has tried to find a new business model. The de-convergence strategy, starting with Time Warner and AT&T, has been a 'must take and follow' business strategy for many communication companies because these are behemoths that dictate the rules and models of business. Several mega communication conglomerates have pursued the new business strategy in order to regain revenues and profits. De-convergence is a new business model emphasizing core business instead of pursuing synergies through the integration of old and new media sectors.

In sum, convergence is not going to disappear anytime soon because demands for information and entertainment, as well as the Internet, remain high (Dennis, 2003). The fall of convergence does not say that the dominant paradigm is dead or will be dead soon because many media companies still seek consolidation. Although it is not conclusive, what we are witnessing is that de-convergence has rapidly become another core business trend, competing with convergence strategies. Media convergence as a form of M&A is still strong, but de-convergence will increase. Media convergence is an ongoing process; however, de-convergence as a rapidly emerging new media model will greatly change the communication market and its ownership structures.

9 Convergence Versus De-convergence in News and Journalism

The newspaper industry has undergone a dramatic shift over the last several decades. The newspaper industry has been one of the primary components of global communication industries and has taken a key role as the major news source for society. With the rapid growth of audio-visual media since the late 20th century, however, the newspaper industry has confronted several challenges from new media, such as television, the Internet, and social media, and it has had to change the traditional distribution of its content through convergence with other media in order to survive, while securing a new revenue resource and new outlet. The changing media ecology has especially forced newspaper corporations to pursue two consequent strategic transformations—the convergence and de-convergence of the newspaper industry and newsrooms—which are some of the most important in the digital media era. On one hand, the newspaper industry has been active in the deal market because many newspaper corporations in several countries have developed their convergence and de-convergence strategies amid a changing media environment. As the media environment surrounding newspaper has shifted, the newspaper industry has had no choice but to admit the necessity of restructuring the industry. On the other hand, the news gathering and distribution process is undergoing deep changes. Convergence and de-convergence of the newsroom in the middle of the rapid growth of digital technologies and broadcasting are major concerns in journalism practices.

This chapter historicizes some of the key features of the global newspaper industry that has been reorganizing since the mid-1980s. It especially maps out the changing structure of the newspaper industry by examining convergence. It then discusses convergence in newsrooms in the digital media era in order to analyze whether newspapers are benefiting from these new trends. Throughout the discussion of the changing business models and new ways of creating and delivering news in the 21st century, it examines what the impacts of digitization and digital delivery on the news value chain are. Finally, it attempts to discuss the recent development of de-convergence in both corporations and newsrooms in the newspaper industry, so that we understand not only the old paradigm of convergence but also the new paradigm of de-convergence occurring in the press sector.

NEWSPAPER JOURNALISM: FROM PUBLIC SPHERE TO COMMERCIAL INSTITUTIONS

Newspapers are the oldest news medium in our modern society. The first regularly published newspaper in the world was published between 1605 and 1609. Rising literacy and philosophical traditions, the formation of nation-states, and a developing postal system all created new market elements in the 18th century and helped newspapers to emerge (OECD, 2010b). The newspaper industry has historically focused primarily on the public sphere instead of being part of the capitalist system (Habermas, 1989). The newspaper industry was originally designed to avoid being directed by the profit motive system. "Independent journalism and news distribution play a central role in informing citizens, because they are a pillar of public life and pluralistic, democratic societies. At their best, they are a source of reliable, quality information that people trust and understand" (OECD, 2010a, p. 17). Of course, newspapers have been the most significant entertainment tool and the most reliable news distributors since their inception as well.

However, with the rapid growth of the capitalist system, it is difficult to be a nonprofit-driven medium, and political-economic approaches to the newspaper industry tend to focus on the political implications of the media's economic structures. That is, the newspaper industry is controlled both by economic processes and structures (Lee, 1998, p. 58). In fact, the newspaper industry is a blueprint of a nature of old economy facing the challenges associated with the emergence of new information and communication technologies, in particular broadcasting and the Internet (Muehlfeld et al., 2007, p. 107). Since the invention of the printing press in the 15th century, the newspaper industry has enjoyed a long economic boom up to the early 21st century.

By 1920, however, newspaper industries began to face major challenges from broadcast radio. For the first time, newspaper publishers were forced to reevaluate their role as primary information providers. Between 1935 and 1936, television was introduced as a news medium, which took off in the 1950s in the U.S., and, later, in many other countries. With the rise of television, the newspaper business faced another major transformation. In the 1960s, television surpassed newspapers as a source of information, and TV networks became more adept at capturing national advertising. Thereafter, the newspaper sector consolidated as family-owned papers were bought by growing chains (Kirchhoff, 2010).

Since the 1970s, the introduction of more television channels and information media has led to a fragmentation of audiences into smaller segments. Until recently, this media development has been accompanied by a steady increase in print-related advertising revenues. People began to think that newspapers were gradually disappearing in the digital era, especially with the rapid growth of the Internet. Since the 1970s and 1980s, some OECD countries have seen significant ownership changes among newspapers (i.e., large entities or media conglomerates incorporating newspapers and the consequent move away from single newspaper ownership). In particular, in the U.S., entities owning newspapers were increasingly listed on the stock exchange, fundamentally altering financial expectations, priorities, and newspaper management due to a greater focus on profitability (OECD, 2010b). In general, newspapers appear to be facing a downward spiral (Muehlfeld et al., 2007). Between 1960 and 1980, 57 newspaper owners sold their properties to Gannett Co. in the U.S. By 1977, 170 newspaper groups owned two thirds of the country's 1,700 daily papers. From 1969 to 1973, 10 newspaper companies went public, including the Washington Post Co., New York Times Co., and Times Mirror Co. (Kirchhoff, 2010). In recent years, several key elements in the newspaper industry, such as the number of physical newspaper titles, their circulation, and newspaper readership, have been in decline in many countries. In the newspaper industry, after a period of healthy growth, newspaper circulations, readership numbers, and advertising revenues are mostly falling (OECD, 2010b, p. 12).

Of course, that does not mean that newspapers are permanently disappearing. Newspapers remain and continue to influence our society, and reading a newspaper is still an important part of our daily activities, although the newspaper industry has become less significant in competition with other media. Contrary to conventional wisdom, newspapers indeed remain a large and thriving industry worldwide, despite the impact of the global recession and the rise of digital media. According to World Press Trends (World Association of Newspapers and News Publishers, 2011; 2012), 2.5 billion people read a daily newspaper at least once a week in 2011, up from 1.7 billion people with 12,477 newspapers worldwide as of December 2009. The number of newspaper companies increased by 1.7% from the previous year. Although daily newspaper circulation fell 0.8% in 2009, it is a relatively small decline, given the depth of the recession. Circulation declines largely occurred in mature media markets of the developed world, including the U.S. and the U.K., while Asia (particularly, China and India) continues to enjoy significant growth: 13% over five years (World Association of Newspapers and News Publishers, 2010).

When we compare different countries, paid circulation in the U.S. has been declining for years and continued to fall in 2008, dropping by 4.6%. Increased newspaper Web site traffic has in part come at the expense of reduced print circulation. Newspapers also are affected by demographic trends because the incidence of newspaper readership among younger people is much lower than among older people. In 2004, daily newspaper circulation was recorded at 54.6 million in the U.S.; however, it dropped to 48.3 million in 2008. It is projected to be 41.1 million in 2013 (PricewaterhouseCoopers, 2009). Similarly, U.K. newspapers have suffered the most dramatic circulation decline of any country outside the U.S. over the last 10 years. Several developed countries, including Japan and Germany, have experienced a readership decline, and they all share very similar reasons.

The consequence of circulation decline in these countries has resulted in massive job losses. The OECD reports that job losses in the newspaper industry have intensified since 2008, particularly in the U.S., the U.K., the Netherlands, and Spain. From 1997 to 2007, it says, the number of jobs in the U.S. has fallen from 403,355 to 356,943, a fall of 12%. The decline has accelerated drastically since then. About half of OECD countries have experienced drops (OECD, 2010b).

However, a few countries in Asia and Latin America have shown increases in paid newspaper circulation. The Asia-Pacific market has become the highest in the world at 309 million in 2008, an increase from 279 million in 2004. Among these, paid circulation in India especially rose by 11 million from 2004 to 2008, bolstered by hundreds of new titles entering the market. The relaxation of foreign ownership restrictions also contributed to growth in India. Foreign companies can own up to 26% of Indian publications, which attracted foreign investment. In China, paid circulation also increased from 88.1 million in 2004 to 107 million in 2008 (PricewaterhouseCoopers, 2009). Since Brazil has experienced rapid growth in circulation, the economic growth of these countries, as well as increases in the population, play major roles in this regard. Among the top 10 newspaper markets, therefore, only three countries—China, India, and Brazil—have experienced increases in paid circulation based on large populations and emerging economies, while other countries have shown declines in circulation (Table 9.1).

Although traditional newspapers in many mature markets have been losing readership, newspaper companies in those markets are at the forefront

Table 9.1 Top 10 Newspaper Markets (Paid Newspaper Circulation, 2004–2013, in thousands)

	Countries	2004	2005	2006	2007	2008	2013 (prediction)
1	China	88,150	96,600	98,700	102,500	107,000	121,500
2	India	74,000	76,000	79,000	83,000	85,000	100,000
3	Japan	70,620	69,700	68,100	68,435	68,000	64,300
4	U.S.	54,626	51,345	52,329	50,472	48,390	40,000
5	Germany	28,193	27,403	26,960	26,452	25,820	23,700
6	Russia	23,265	22,750	22,600	22,500	22,400	21,900
7	U.K.	17,020	17,106	16,650	16,074	15,500	13,450
8	South Korea	14,610	14,580	14,565	14,415	14,300	13,750
9	Brazil	6,275	6,532	6,955	7,775	9,320	10,200
10	France	7,900	7,775	7,650	7,613	7,575	7,320

Source: PricewaterhouseCoopers. (2009). *Global Entertainment and Media Outlook 2009–2013*, 465–503.

of the digital revolution. Many newspaper companies in mature markets have in fact embraced digital platforms and new forms of print publishing, growing their product portfolios, audience reach, and revenues, even while their traditional print circulations have come under pressure (World Association of Newspapers and News Publishers, 2010). This means that many newspaper corporations have pursued a print-based but digitally expansionist new media business model, which is both a problem and an opportunity for the industry. In fact, reading news online is an increasingly important Internet activity. For some countries, online readership figures have overtaken the number of offline readers. In some OECD countries, more than half of the population read newspapers online (up to 77% in Korea), but at least 20% of the population read online newspapers. In many OECD countries, TV and newspapers are the most important sources of news, but this is shifting with newspapers losing ground more quickly to the Internet than TV. In countries such as Korea, the Internet has already overtaken other forms of news. For the most part, reading news online complements other forms of news reading. Most surveys show that active offline newspaper readers tend to read more news online. Countries where offline newspaper reading is less popular than online newspaper reading are the exception (OECD, 2010b, p. 9).

The current changes in the newspaper industry have led to several significant shifts in the media industries. Most of all, the rise of the Internet and other digital technologies radically changes how news is produced and diffused. They enable the entry of new intermediaries that create and distribute news, including online news aggregators, online news publishers, mobile news actors, citizen journalism, and many others (OECD, 2010b). The new situation also challenges the existing U.S./U.K. duopoly in the newspaper industry. Until the late 20th country, these two countries seemed to dominate global newspaper markets, as well as magazine markets. Though several Japanese and Chinese newspapers figured among the world's top 10 newspapers in terms of circulation, they were rarely read outside their countries of origin. In contrast, the Anglo-American press has global reach and influence. Some publications, such as the British weekly news magazine *The Economist*, even sell more copies outside Britain (Thussu, 2006, p. 123). In fact, several newspapers and magazines in the U.S. and the U.K., including the *Wall Street Journal*, *Business Week*, the *Economist*, and the *Financial Times*, have led the file of global business journalism. However, with the decline in circulation of these major papers headquartered in the U.S. and the U.K., the dominant role of these countries seems to lose ground in the global press industry as well. Several newspaper companies in mostly Western countries, however, have expanded their size and market dominance through media convergence. One after another, newspaper companies have acquired other small and mid-sized newspaper companies both globally and domestically, so the dominant role of Western newspapers is not yet fading out.

RESTRUCTURING OF THE NEWSPAPER INDUSTRY

The global newspaper publishing market (defined as online and offline circulation and the advertising revenues of traditional newspaper publishers) was estimated at a worth of $157.2 billion in 2010. Despite the fact that 2010 was a year of decline, its revenues considerably exceed those of recorded music ($26.2 billion), video games ($58.3 billion), and films/movies ($87.1) (PricewaterhouseCoopers, 2009). In the midst of the rapid growth of digital media, newspaper companies have attempted to survive and defend their status as a leading news medium. What they have strategically pursued has been a corporate diversification strategy in the audio-visual industry and the development of online editions. However, companies with a multimedia business portfolio face important challenges in order to be more efficient in their new businesses, such as finding possible synergies as a result of newsroom convergence (García Avilés & Carvajal, 2008). Therefore, multimedia diversification strategy, platform (newsroom) convergence, and online development are among the most important responses adopted by newspaper corporations. These responses, of course, have, to a large extent, been triggered by technological convergence and the relaxation of market regulation. In this section, I focus on integration within the newspaper industry and with other audio-visual sectors, while emphasizing platform convergence in the next section as major parts of the structural transformation of the newspaper industry.

As expected, the newspaper industry is not an exception within neoliberal transnationalization, although the degree of the deals has been the lowest among communication industries as a reflection of its unique characteristic emphasizing domestic readers. Since newspapers are based on written communications, the majority of readers in most countries are those who read the native languages. While the trend toward integration in the U.S. newspaper industry goes back to the late 19th and early 20th centuries, the major activities started with the rapid growth of neoliberal regimes in many countries, as in other media industries. Although the corporate environment of the 1970s and 1980s was relatively stable, it has become difficult for newspaper firms since the 1990s. As a result, facilitated by the relaxation of the regulatory climate, M&As have become an important strategic tool for firms in the newspaper industry, as in many other media industries (Chon et al., 2003).

Over the period 1982–2009, the number of M&As completed only in the newspaper industry worldwide rapidly increased. Of course, according to previous research, a large number of initiated M&As in the newspaper industry are abandoned even before completion. Since the newspaper industry is a declining industry prone to consolidation, such acquisitions may be even more likely to fall prey to shareholder concerns, litigations, and unexpected disagreements on financial terms (Pickering, 1983). Regulatory roadblocks constitute another important cause of termination, especially for horizontal mergers. Not surprisingly, the abandonment rate for the newspaper industry is even higher at 27.2% (Weston et al., 2004). Unlike previous works, this chapter only analyzes M&As completed without consideration of original attempts.

The overall number of M&As in the newspaper industry, which is horizontal integration among newspaper corporations, was recorded at $320.7 billion with 2,087 deals between 1982 and 2009. In considering the distribution of deals in the newspaper industry over time, it is clear that M&As in the newspaper industry consistently increased until the mid-1990s and substantively soared between the mid-1990s and the early 21st century. The number of deals in the newspaper sector was only 13 in 1982 and 31 in 1983. The situation continued until 1994; however, the number of deals has increased since 1995 when the number recorded was 107. Several large players purchased small newspaper companies; however, there were also a number of smaller players in the deal market. This data only explains corporate convergence between newspaper firms; but some of these newspaper companies also own broadcasting companies, such as cable and terrestrial—therefore, the nature of the transactions in the newspaper industry may naturally show vertical integration as well. Unlike other communication industries, though, the deals were not significant in terms of either the numbers or transaction values. In 2000, the market peaked at only 130 deals. Due to the financial crisis, however, the number of deals in the newspaper industry plummeted to 49 in 2003. After its recovery, the number of deals was again reduced from 110 in 2006 to 80 in 2009 (Figure 9.1).

During the same period, the U.S. was the largest country in the deal market, and it acquired 956 companies (45.8%), followed by the U.K. (167 cases), Australia (134 cases), and Germany (120 cases). These four countries

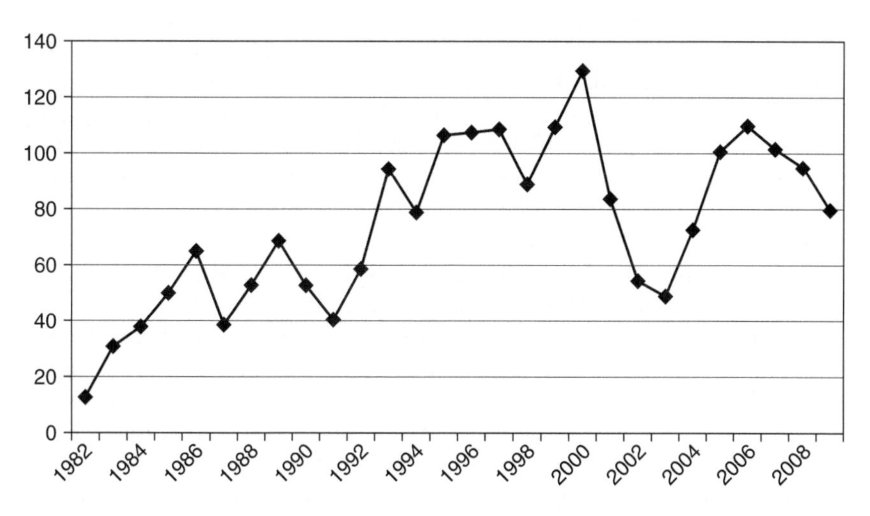

Figure 9.1 M&As in the Newspaper Industry, 1982–2009 (millions of dollars)

together accounted for 1,377 deals (66%). The role of U.S. companies in the newspaper market as major players has not changed much during the period between 1982 and 2009. The relatively larger role of Australia seems obvious mainly because of Rupert Murdoch's continuous interest in newspaper. Since News Corporation started as a newspaper corporation, Murdoch has tried to purchase newspaper companies in the U.S., the U.K., and Australia. For example, News Corporation purchased the financially struggling *Wall Street Journal* in 2007.

M&A activities in the newspaper industry have resulted in the concentration of ownership and a change of the major players in the market. In 2006, for example, Knight Ridder, the second-largest newspaper company in the U.S., had to sell itself to McClatchy, a newspaper publisher less than half Knight Ridder's size. McClatchy acquired 20 of Knight Ridder's 32 daily newspapers, including the *Miami Herald* and the *Kansas City Star*; McClathy paid $4.5 billion in cash and stock to Knight Ridder, which was struggling due to the decline in readership and advertising revenue (Lieberman, 2006). As is well known, two major reasons support the massive M&As in the newspaper industry. One is the development of a broad and strong capital market and the existence of institutions that utilize M&A activities to expand their capital gain. The other is the elimination of competition in the newspaper market since the early 20th century (Adams, 1995). Top-tier newspaper chains certainly understand these reasons and act accordingly.

The situations are not much different in other countries. The concentration of chain and conglomerate ownership in Canada during the late 1990s (about 95% of newspapers were controlled by six chains) was far in excess of the concentration in the U.S., where the largest 15 chains owned 25% of the daily newspapers (Soderlund & Romanow, 2005). In China, newspaper groups have also ventured into the capital market. *Guangzhou Daily* became the first newspaper group in 1996, and there are currently a total of 49 newspaper groups in China, all of which are undergoing the restructuring of their operational assets. In addition, more than 10 newspaper groups made preparations to go public in 2009 (Liao, 2009). In the case of China, news media have become liberalized and commercialized. At the macro level, the purpose of the Chinese media was exclusively ideological propaganda in the old days, but now, with more economic forces influencing the media sectors, the attention of policymakers is shifting to how to construct a more rationalized and market-oriented media system (Zhang, 2010).

Meanwhile, there were 482 cross-border deals during the period 1982–2009. Unlike other communication industries such as broadcasting, advertising, and telecommunications, which have been discussed in detail earlier, in the newspaper industry, cross-border deals were not popular. While the U.S. (66 deals), Germany (66 deals), Australia (63 deals), and the U.K. (62 deals) together account for 53.3% of cross-border deals, the U.S. consisted of only 13.6% in this category. As these data prove, transnationalization in the newspaper industry is unique. Unlike other communication industries, the major role of the U.S. in cross-border deals has not been a powerful influence in the global M&A market. The proportion of U.S. deals in cross-border dealing is relatively less than in other communication sectors, such as broadcasting, advertising, and telecommunications. As Herbert Schiller pointed out, the newspaper industry was traditionally one medium of exception to the transnationalization in many industries. "For newspapers, it is the national and, even more, the local/regional market that is determining" (Schiller, 1989, p. 36). While the newspaper industry has been part of the global M&A market, it is the least significant in terms of the number of deals. Interestingly, the newspaper industry has shown to have been less influenced by the two major events (1996 Telecommunications Act and the 1977 WTO agreements) than either telecommunications or broadcasting industries. Although the number of cross-border deals in the newspaper industry has increased since the late 1990s, their numbers are not substantial, and the cross-border deals in newspaper were more active in the 1980s, unlike in other communication industries.

PLATFORM CONVERGENCE AND DE-CONVERGENCE

Cross-platform convergence in newsrooms, spurred by technological innovation and mainstream journalism's desire to reverse audience declines, has been one of the most noticeable changes in journalism over the past decade (Thornton & Keith, 2009; García Avilés & Carvajal, 2008). In the early 21st century, print-broadcast content-sharing partnerships were called inevitable, and a model for the future. Convergence in journalism is what takes place in the newsroom as editorial staff members work together to produce multiple products for multiple platforms to reach a mass audience with interactive content, often on a 24/7 timescale (Pryor, 2004, as cited in Melinda, 2011). When the Chicago-based Tribune Co., owning the *Chicago Tribune*, 22 television stations, and four radio stations, as well as cable and Internet holdings, bought Times Mirror, owning the *LA Times*, in 2000, several commentators argued that this acquisition style is a compulsory path for the newspaper industry because newspaper and broadcasting companies can share information. Once separated, these two major news outlets are now under same umbrella; therefore, they have newspaper reporters appearing on television, while TV reporters write stories for the newspaper. They believe that this kind of convergence is going to continue in the traditional media, and part of it is feeding the Internet (NewsHour, 2000). Television became a coveted partner of newspapers, and executives talked of synergy between the two media (Ahrens, 2007).

Many newspapers have also pursued convergence with the Internet. For example, *Vocento*, one of the largest media groups in Spain, has expanded its size through the acquisition of several newspapers, four television stations, and three radio stations, together with online companies. The historical strength of

Vocento was regional newspapers, but the company has pursued an aggressive strategy of diversification toward audio-visual and online media. *Vocento* was the first Spanish media group that implemented a regional multimedia strategy as its core business in 2001. Following *Vocento*, many Spanish newspaper publishers have developed a multimedia strategy to gain several competitive advantages. Several Spanish newspaper companies also developed their strategies in the Internet market, because newspapers' online editions are suitable platforms to distribute multimedia content and the best area to generate content, advertising, and marketing synergies. Until 2005, publishers limited themselves to only publishing the contents of their print editions on their websites, and they were hardly innovative. Currently, their cross-media strategy provides their online editions with multimedia packages, online chats, blogs, exclusive online content, video reports, and other recourses (García Avilés & Carvajal, 2008, pp. 458–459). As of March 5, 2012, about 224 newspaper publishers were listed for sale in the Merger Network database (2012), the online community for deal market. However, news organizations are no longer embracing the print-broadcast shared-newsroom model. Some refused to jump on the convergence bandwagon; others have abandoned partnerships outright; and still others reported less-than-enthusiastic attitudes toward existing partnerships (Thornton & Keith, 2009). Newspaper and TV station owner Belo Corp. also finished the spin-off of its newspaper division to create separate newspaper and television companies due to financial reasons in 2008 (Belo, 2008). Belo, which keeps 20 television stations and their websites, said that the spin-off was triggered by major changes in the media industry, which have made print and broadcast assets a focus for divergent investor groups. This means that Belo was not considering the role of the media as a social institution, but considering it as a solely commercial agency to give investors a greater insight into two distinct businesses (Macmillian, 2007).

In the U.S., in January 2007, the New York Times Co. sold its nine TV stations for nearly $600 million, months after it left its partnership with Discovery Communications on a joint TV channel. Tribune Co., which had 11 newspapers and 24 broadcast stations and has put itself up for sale, has found tepid interest from bidders to buy the company and is considering spinning off its TV stations. An executive at E.W. Scripps, which owns 19 newspapers and 10 TV stations, said it might consider splitting off its newspaper division as a separate company (Ahrens, 2007). When Tribune Co. bought Times Mirror in 2000, the new company had newspapers and TV stations in several large markets. The company lobbied for and anticipated a repeal of the Federal Communications Commission's ban on one company owning a newspaper and TV station in the same city and looked to expand its holdings. The ban is still in place, and the expected advertising synergies have not been achieved. Even though terrestrial TV stations are still profitable, they no longer enjoy the dominance they did in the days before cable and the Internet. And in many places, the newspaper and television cultures never meshed. It was a failed experiment for many media companies (Ahrens, 2007).

The newspaper business does not need to have expensive TV cameras. In order to post their videos online, they need a small video camera, or they hire entry-level videographers to shoot digital video of a news event and post it on a website rather than pay a TV reporter and producer to create a three-minute news report for television. Many newspapers are training their reporters to shoot video with their stories. Some reporters carry video cameras and shoot video to accompany their articles when they appear on their newspaper's website (Thornton & Keith, 2009; Ahrens, 2007). As digital technologies have driven convergence between newspaper firms and broadcasting companies, the same digital technologies have caused a recent de-convergence trend at many newspaper companies.

The New York Times Co. also sold its Regional Media Group to Halifax Media Holdings, of Daytona Beach, Florida, for $143 million in cash in December 2011. The New York Times Co.'s regional publications have been hit by declining revenues. From 2008 to 2009, ad revenue fell 30.2% and declined a further 8.2% in 2010 (Vega, 2011). Regional newspapers have struggled recently because of weak local retail and national advertising, partly reflecting the economy's broader travails. In July 2011, the New York Times Co. sold more than half of its stake in the Fenway Sports Group, the owner of the Boston Red Sox, for $117 million. The sale of such assets allows the company to focus on its anchor newspapers, the *New York Times*, the *Boston Globe*, and the *International Herald Tribune* (Chozick, 2011).

Newspaper corporations have undergone a transition from a convergence model to a de-convergence model; however, the important thing is that they have changed and are still changing their business models, not because of the social values of newspapers, but because of their commercial values. Journalism has been regarded as a public service by all of the commercial media. Historically, journalism is something that newspapers, broadcasters, and magazines regarded as an activity directed toward noncommercial aims that are fundamental to a democracy—aims that could not be bought and sold by powerful interests.

However, some of the corporate media owners maintain their journalism holdings not merely to make profit but also to promote their probusiness, antilabor view of the world (McChesney, 1999). Economic pressures are becoming the primary forces shaping the behavior of newspaper companies (Picard, 2004, p. 54). The rapid involvement of news journalism in the capitalist system has resulted in the decline and marginalization of public service values among newspapers, placing the status of nonmarket public service in jeopardy across society (McChesney, 2008; 1999). It is increasingly clear that some newspaper owners are affecting journalistic quality, producing practices that diminish the social value of newspaper content and that divert the attention of newspaper personnel from journalism to activities primarily related to the business interests of the press (Picard, 2004, p. 540).

In fact, newspapers, which are symbols of a public service, have been major commercial institutions to be sold and bought in order to make profits, instead of serving the interests of the enhancement of democracy (McChesney, 2008; Picard, 2004). They have tried to make profits through both commercialization and commodification, not through newspapers' value as public service institutions. As Mosco clearly states, "commodification refers to the process of turning use values into exchange values, of transforming process whose value is determined by their ability to meet individual and social needs into products whose value is set by their market price" (2009, p. 132). Commodification is what has been happening as the global newspaper industry attempts to benefit in a capitalist market system (Lee, 1998). The consequence of the commodification of the newspaper industry is the diminishing role of the newspaper as a public sphere medium, and the newspaper industry faces serious problems because it has lost both social values and commercial values at the same time.

CONCLUSION

This chapter has analyzed the structural transformation of the newspaper industry as a form of convergence and de-convergence in both newspaper companies and newsrooms. The newspaper industry had been a symbol of journalism and newspapers were considered to be a social institution perpetuating the public sphere instead of a part of the capitalist system. However, the introduction of neoliberal economic and communication policies have shifted the global newspaper industry from a social institution to a profit-driven commercial sector. Since the mid-1980s, the newspaper industry has continuously pursued new strategies in order to survive in the midst of competition with broadcasting and later digital media. The solutions they have actualized are the diversification of their companies into broadcasting and new media, corporate convergence through both vertical and horizontal integration, and convergence in the newsroom.

While new technologies have been major targets of media corporations, the newspaper industry has to compete primarily with new media, which is a disadvantage for newspaper companies. For newspaper companies, therefore, M&As have been very strong tools to becoming media giants, although the same method has suffocated their corporations, which has resulted in a change of business strategies, from convergence to de-convergence. While TNCs, in both media and nonmedia sectors, have played a significant role in the process of change in many media industries, newspapers are relatively small potatoes in comparison. Through convergence strategies, these Western-based media TNCs have become media giants, and the emergence of mega broadcasting corporations has allowed large companies to control both content and channels to maximize their dominant position.

However, the newspaper industry has shown signs of strain in implementing convergence. It has suffered from M&As, in particular with broadcasting corporations. This also means that newsroom convergence between the newspaper and broadcasting sectors has not fulfilled forecasted synergy effects. During the financial crisis in the early 21st century, newspaper corporations have lost their revenues from advertising as well. Therefore, they have no choice but to pursue de-convergence in both corporations and newsrooms. While the convergence of the newspaper industry still remains a powerful force, after the failure of several merged newspaper corporations, such as Tribune Co., New York Times Co., and Belo in the U.S., many broadcasting mega giants in the world are pursuing a new direction. Economic realities have certainly played a pivotal role in determining the future of the newspaper industry.

As for convergence in the newsroom, the growing dominance of the Internet as a source of information of all kinds will continue to affect news journalism. Convergence has changed the way in which news is made. Digitization and technological convergence mean that the boundaries of media platforms are easier to cross. Content can easily be shared between journalists making news for television, radio, and the Internet. Media organizations increasingly integrate production for different media platforms in order to encourage cooperation between desks (Erdal, 2007). While broadcasting and newspapers are de-converging, this implies that newspaper and online journalism will continuously converge. The media economy surrounding the newspaper industry hinges on the Internet, not the newspaper industry per se. The newspaper industry has electronically connected with the Internet, which means that one of the oldest media is now part of the new media. The reality is not always positive, and newspaper companies try to keep their own unique boundaries while changing their content to attract visually driven young audiences. Newspapers are confronting an acute dilemma, juggling how to give citizens more of what they need in a democracy and meeting commercial demand in the marketplace (Entman, 2010), which is an unprecedented challenge. However, it is crucial to strive for a reasonable balance between these two aspects, which means that newspapers must consider the major role of journalism for democracy, while pursuing its structural transformation, whether convergence or de-convergence, because newspaper journalism still remains the core institution for the public sphere.

10 Conclusion

The communication industries have become the center of our contemporary politics, economy, and culture. Once crucial for the establishment of the nation-state and democracy, the communication industries have played a key role in the rise of capitalist commercial enterprises. As McChesney (2001) points out, the current era in history is generally characterized as one of globalization, technological revolution, and democratization. In all three of these areas, communication plays a major role. Economic and cultural globalization arguably would be impossible without a global commercial media system to promote global markets and to encourage consumer values. Media also function as a public sphere to secure and develop democracy in many countries. Most of all, the global communication system, as part of global capitalism, has developed commercial values, although we cannot deny the central role of media and communication in democratization.

The move toward the commercialization and commodification of the communication apparatus has been noticeable since the early 1980s. The global communication industries, both media and telecommunications, have pursued corporate convergence in order to control the whole media sector, from the production to the distribution of content through digital operating systems, as well as to create valuable commodities in the media markets. They have sought synergy effects, economies of scale, and the production and distribution of diverse cultural products to maximize their profits amidst neoliberal globalization and the growth of digital technologies. Through vertical and horizontal integration, these corporations have rapidly expanded their scales in the global markets.

Two primary factors—neoliberal globalization and digital technologies—have played a major role in the restructuring of the global communication industries. On one hand, as neoliberal globalization evolves, transnationalization of the communication industries has occurred daily, which means that many communication corporations in Western countries have massively invested in non-Western countries as a form of M&As and joint ventures. These media TNCs have targeted emerging markets, such as China, India, Brazil, and Korea, in order to benefit from their emerging economies. The trend toward the transnationalization of the communication sector through

M&As around the world indeed began with the rise of neoliberal globalization in the early 1980s, and it has further accelerated with the 1996 Telecommunications Act and the 1997 WTO agreements. A politics of neoliberal media reform has driven many countries to restructure their communication systems as a major part of economic transformation. The changes that have been made in communication market structures indicate that competition, once the exception in communications, was quickly becoming the norm. As a result of a series of deregulating markets, transnational capitals have been active in both domestic and global communication markets.

On the other hand, the emergence of the communication industries in the early 21st century has been phenomenal due to the rapid growth of the Internet industry. With the swift growth of technologies, our contemporary society has witnessed a dramatic change. While traditional material-based industries, such as auto, furniture, and steel industries, are still significant in national and global economy, information and communication technologies (ICTs) have rapidly become a new area that corporations desire to own and control. Due to the significant role of ICTs, in particular the Internet and now social media, such as social networking sites and user-generated content (e.g., YouTube) technologies, many communication firms have integrated the Internet services sector. These digital technologies have enabled the media and telecommunications industries to integrate with other industries. In the digital communication era, the Internet is a content distribution platform and a more effective means of transmitting voice, data, and video than proprietary networks; therefore, many communication companies have aggressively expanded their business areas through integration with Internet-related firms (Rainie & Wellman, 2012; Chambers & Howard, 2005; Liu & Chan-Olmsted, 2002). Consequently, convergence between content providers, such as television and radio networks, and transmission channels, such as those dealing with Internet service networks, has become most important in the media and telecommunications industries (Jenkins, 2006), and digital technologies and neoliberal communication policies have played major roles in the convergence process. These two elements work in a complicated way, which has resulted in the change of ownership in the Internet services industries.

Meanwhile, it is also crucial to understand that the transition of the global media industries can take place in the context of the increasing role of financialization in conjunction with neoliberal globalization. As Winseck emphasizes (2010, pp. 366–367), financialization has directed our attention to the capitalization of the media industries alongside the traditional focus of critical media political economy on media ownership, markets, regulation, commodification, digitization, and so on. The concept highlights the extraordinary growth in the size of the financial sector and financial assets relative to the industrial and other sectors of the economy over the past 25 years and especially since the mid-1990s. These developments have been enabled by the steady liberalization of financial markets; the search for new modalities of capital accumulation in the face of persistently low levels of overall economic growth in the Western capitalist economies since the 1970s; the rapid growth of network information and communication technologies; and accelerated global flows of capital.

The logic of financialization is especially important to recent developments across the communication industries because it has, paradoxically, created greater media concentration but also bloated media giants that have sometimes stumbled badly and occasionally been brought to their knees by the two global financial crises of the 21st century (Winseck, 2010). Indeed, several bastions of the 'old order' that assembled just before or after the turn of the millennium have subsequently been restructured (Bertelsmann, ITV) or dismantled (AT&T, Vivendi); have collapsed in financial ruin (Canwest, Craig, Kirch); or have abandoned early visions of convergence altogether (Bell Globemedia, Time Warner). The woes of these entities offer a cautionary tale regarding the impact of financialization on the media, rather than a tale in which the Internet, changing media behaviors, and declining advertising have precipitated a "crisis of the media." These trends are global in scope and are ongoing (Scherer, 2010).

We cannot deny the fact that neoliberal globalization in tandem with financialization has greatly influenced the transformation of the global communication industries. In the 1990s and until the early 21st century, neoliberalism, globalization, and financialization have been converging to find synergy effects through the globalized integration of the communication industries.

However, promised synergies could not be fulfilled in the midst of several financial crises during the periods of 2000–2001and 2007–2008, and during the crisis of European countries in 2012, because corporations and countries are too closely connected. Once one country or one company experiences crisis, it triggers other countries or other corporations. Interconnectivity and integration become culprits of the crisis in many occasions, instead of driving the growth of the global economy and global media industries. Digital technologies and globalization have become two major drivers for the convergence paradigm; however, they are also providing opportunities for global media corporations to de-converge their conglomerated firms in the early 21st century.

Consequently, the communication industries have been structurally altered since the early 21st century, because major communication corporations have adopted de-convergence in the midst of the failure of media convergence. While M&As among communication companies have been the major trend since the late 20th century, many communication corporations have begun to focus on a few core businesses while conducing de-convergence strategies. Mega communication corporations in many countries, not only in the West but also in the East, have begun to sell parts of their companies and split off and/or spin off their firms rather than consolidating their corporations. From mega communication giants, such as CBS-Viacom, NBC

Universal, Vivendi, and AT&T to small local media corporations, they have experienced serious setbacks after corporation integration, and they have strategically pursued de-convergence. Media firms have had to reshape their dysfunctional organizational structures resulting from previous corporate convergence, and their solution has been de-convergence, which divests the media units that do not fit well with other parts of their corporation (Alexander & Owers, 2009). Media and communication corporations have to adopt new business strategies to survive, and the solutions they take are based on a de-convergence paradigm.

The trend of media de-convergence has become a new watchword. As convergence emerged as a buzzword in the 1990s, de-convergence has emerged as a central paradigm in the communication industries in the early 21st century. De-convergence has been mainly seen over the past decade as a reaction to the well-documented business failures of convergence. The move signified that two of the most popular early convergence applications—corporate convergence and newspaper/television convergence—had proved misguided (Edge, 2010). De-convergence, again, refers to the business activities in which media corporations strategically decrease the magnitude of their companies in order to regain companies' profits and public images and/or to survive in the market with several business strategies, either by selling parts of the shares to other companies or by splitting off and/or spinning off of their companies. De-convergence has swiftly replaced convergence in the communication industries. Mega communication corporations have begun to find new business models after the loss of their revenues and profits with convergence. While convergence is still a powerful corporate policy, many media corporations have strategically turned their interests toward de-convergence. De-convergence has indeed become one of the major business paradigms in the global communication industries in the early 21st century.

Convergence is not disappearing, of course. Google acquired Motorola Mobility for $12.5 billion in 2012 because of Motorola Mobility's full commitment to the Android operating system, and thus selective convergence will continue (Google, 2012). As Facebook, the most famous social networking site in the world, started to raise money in the stock market, creating a $104 billion media giant on May 17, 2012 (Rusli & Eavis, 2012), financialization continues. Likewise, social media, including SNSs and smartphones, have become new commodities, and they might be new engines that media corporations must acquire to survive. For example, only a few years ago, Samsung Electronics Inc. was considered as a semiconductor maker; however, as of May 2012, people now know the company as the largest smartphone maker in the world. In the SNS market, due to the phenomenal growth of Facebook, Twitter, and other SNSs, media giants hope to acquire at least parts of these corporations. This means that we might expect to see another major tide of media convergence driven by these social media. During the period 2005–2012, these areas merged with one another and created the converged communication platforms (Mulligan, 2012, p. 306).

Unlike previous convergence between old and new technologies, however, new technology versus new technology-driven convergence might be a new norm, while de-convergence aiming at separating the old and the new sectors is going to be burgeoning. Convergence and de-convergence are not mutually exclusive, and de-convergence cannot entirely take over the convergence trend due to the continuation of media convergence. What we have to understand is that de-convergence has become another major paradigm in the communication sector.

In sum, the global communication industries will continue to be influenced by the changing socio-political-economic milieu in the near future as they have been over the last few decades. Therefore, instead of emphasizing only one aspect of the global communication industries, we need to further analyze the changing patterns of the global communication industries within the broader socioeconomic environment into which the global communication industries are integrating. Johnson et al. clearly states that "[h]istory is not about the past but about the relationship between the past and the present" (cited in Nerone, 2011). Communication historians and anyone who is interested in media and communication need to accept as common sense the relationship between past paradigm and present paradigm in the communication system (Nerone, 2011, p. 7), which will eventually lead us to contemplate our future in the realm of media and communication.

Notes

NOTES TO CHAPTER 1

1. Previous work focused on only a few years' changes and one specific industry with few exceptions. For example, Herbert Howard (1998) examined the effect of the 1996 Telecommunications Act on the ownership of television stations in the U.S. based on data between 1996 and 1997; therefore, he did not analyze the long-term influence of the Act. Sylvia Chan-Olmsted (1998) investigated the M&A strategies for broadcasting, cable TV, and telephone companies before and after the 1996 deregulation; however, the analysis did not provide the new trend starting in the 21st century, so it showed a limited interpretation of the effect of the 1996 Telecommunications Act as well. In addition, these previous studies have not focused on the initial impact of the neoliberal reform initiated by the U.K. and the U.S. in the 1980s because they used data mainly from the 1990s.
2. Lawson-Borders (2003) defined convergence as the blending of old media (e.g., traditional media such as magazines, newspapers, television, cable, and radio) with new media (computers and the Internet) to deliver content. However, convergence is an elusive term that is used in multiple contexts and is often ambiguous in its definition.
3. Some scholars primarily consider the cross industry M&As as communication convergence; however, their analysis can't represent the overall trend of convergence, since it does not consider M&As within the same industry, as in the case of cable (new) versus network (old), as well as the Internet (new) versus telephony (old). For example, if one needs to consider only cross-industry M&As, one cannot define the merger between Viacom (mainly cable TV) and CBS (mainly network TV) as an example of convergence, which is not the case (Chon et al., 2003).
4. The transformation of the global communication industries has been driven by a few Western countries in the midst of neoliberal globalization. However, with the recent growth of non-Western media and telecommunications industries, many scholars have discussed de-Westernization and de-centering Western culture over the last several years. Although the term 'the West' has brought about many different meanings and phenomena, as in the case of 'the East,' there is no consensus over their respective meanings. According to the *New Oxford Dictionary of English* (2003), the term 'the West' has two different meanings: One is Europe and North America seen in contrast to other civilizations, and the other is the non-Communist states of Europe and North America, contrasted with the former Communist states of Eastern Europe. However, the reality is far from these definitions because the West is

no longer only in Europe, and not all of Europe is in the West (Jin, 2009b). More importantly, the basic idea of the West versus the East has lost its grip amid globalization. It is true that geography has always mattered, and for many, it matters now more than ever. Edward Soja (1998, p. 1), for example, suggests that one sees the formation of new postmodern geographies, and argues that "today it may be space more than time that hides consequences from us, the making of geography more than the making of history that provides the most revealing tactical and theoretical world"; however, at least geography is not a sole factor in deciding global affairs amid globalization due in large part to the integration of the world culturally, economically, and technologically, which means geography reveals much less these days. As Mosco also argues (2004, pp. 118–119), geography already ended in the 1850s when the telegraph was introduced, and it ended again a few decades later when electrification lit up the cities, and again and again and again with new technologies, such as the telephone and cyberspace. Of course, Mosco (2005, p. 53) acknowledges that cultural differences are not easily surmounted, although advances in telecommunications and computing have overcome some geographical differences. So, the underlying premise of this book is that the West is a historical, not a geographical, construct (Jin, 2009b). There is no doubt that the field has long been one-sidedly and hegemonically dominated by Western thoughts, theories, methodologies, and practices (Satoshi, 2007). The globalization of Western media has been a major influence in shaping media cultures internationally, and the globalization of a powerful medium has tended to increase Western cultural influence (Thussu, 2006). One of the critical points that determines the status of a country in its relation to the West and the East is again not geography but several factors indicating the level of modernization, such as its respective level of development, industrialization, urbanization, capitalism, secularism, and modernization (Hall, 1996; Venn & Featherstone, 2006).

5. Detailed reasons for the failure of convergence in terms of technological convergence between television and the Internet could be read in Castells (2001, pp. 189–190).
6. As William Lazonick (2009, pp. 16–17) points out, a business model can be characterized by three components, comprised of (1) its strategy—the types of product markets for which a company competes and the types of production processes through which it generates goods and services for these markets; (2) its finance—the ways in which it funds investments in processes and products until they can generate financial returns; and (3) its organization—the ways in which it elicits skill and effort from its labor force to add value to these investments. While these three major categories are not separate from each other, this book, again, especially emphasizes that financialization is one of the most significant business strategies for making profitable corporations.
7. Spin-off is a type of divestiture, which is the creation of an independent company through the sale or distribution of new shares of an existing business or division of a parent company. Meanwhile, split-off is a type of corporate reorganization whereby the stock of a subsidiary is exchanged for shares in a parent company.

NOTES TO CHAPTER 2

1. The 10 entertainment and media industries include film, broadcast and cable television, cable and DBS distribution, the Internet, magazine publishing, newspaper, book, radio and outdoor advertising, and theme parks and amusement parks. Except for the U.S., other countries' data does not include spend-

ing in theme parks and amusement parks because the book does not provide detailed information for this. See PricewaterhouseCoopers, *Global Entertainment and Media Outlook: 2000–2004* (New York: PricewaterhouseCoopers, 2000), p. 6.
2. Calculated from PricewaterhouseCoopers, *Global Entertainment and Media Outlook: 2000–2004*, pp. 7, 9, 11, 83, 127, 220, 248, 303, 356.
3. The fall of the Berlin Wall in 1989 and the breakup of the Soviet Union two years later transformed the landscape of international politics, profoundly influencing global information and communication. Since then, the communication industries in the eastern bloc countries have gradually been converted to the market-oriented system (Schiller, 1999b, pp. 43–44; Thussu, 2006, pp. 30–31).
4. The privatization and liberalization of telecommunications have also been stimulated by the desire for profits by telecommunications firms and the investments bankers who coordinate the neoliberal reform (Herman & McChesney, 1997, p. 111).

NOTES TO CHAPTER 3

1. Multichannel television refers to services that provide additional programming beyond the free-to-air analogue terrestrial channels. Multichannel TV services should be broken down by CATV, DTH (Direct-to-home TV), IPTV (Internet Protocol TV), and DTT (Digital Terrestrial TV).
2. Among the total 11,055 deals during the period, in the case of 369 deals, the acquirer countries are unknown. Therefore, the data eliminates these unknown cases for documenting cross-border deals.
3. I calculated film and program rental items, which were under the name of 'Other Private Services,' in several years from *Survey of Current Business*.

NOTES TO CHAPTER 4

1. The 1990 and 1999 data came from "Global Ad-spend Trends: World Advertising Expenditure," *International Journal of Advertising* 20(2): 266–267 (2001). The *Journal*, which was published by World Advertising Research Center, recopied this data from *World Advertising Trends 2001*, which was published by the same research center. The 1980 data came from Mike Waterson, "International Advertising Expenditure Statistics," *International Journal of Advertising 11*(2): 18 (1992). The 1980 data included advertising expenditure in the U.S., Europe, and Japan, but Africa, the Middle East, and Latin America accounted for about 3.6% of total global advertising spending in 1990; therefore, the 1980 data would represent the global advertising expenditure in the same year.
2. Unlike the data from 1980 and 1999, which included advertising spending on TV and radio, in print, in cinemas, and outdoors, the 2002 data added advertising spending on the Internet.

NOTES TO CHAPTER 5

1. For example, China's practice is state sponsored, a voluntary rather than an imposed action from imperialists. Although Hollywood barons have long coveted the huge market of China, it is China itself that took the initiative to open the gate to Hollywood (Su, 2010, p. 53).

2. In December 1906, Pathé Frères, by then perhaps the most powerful producer in the world, opened one of the first purpose-built cinemas in Paris. By 1909, Frères had a chain of 200 cinemas in France and Belgium. It was clearly in the interests of producers and distributors to show films first in cinemas that they themselves owned, or at least to exclude competitors' films as much as possible. This was a substantial reason for the creation of the Motion Picture Patents Company (MPPC) in the U.S., set up in 1908 by Thomas Edison and others to exercise control by means of patent enforcement (and, failing that, violence) over film production and exhibition (Terra Media, 2010).
3. While cross-industry M&As as communication convergence are crucial, vertical integration can't represent the overall trend of convergence, since it does not consider M&As within the same industry. Therefore, it is necessary to analyze both vertical and horizontal integrations in the film industries in order to determine the bigger picture of convergence in the film sector.
4. In order to discuss the structural transformation of the film industries based on comprehensive empirical data, the next three sections analyze three major issues: (1) the overall trend of M&As in production (Standard Industrial Classification-based data: SIC 7812), distribution (SIC 7822), and exhibition (SIC 7830) in the number of deals and the total amount of transaction values, if disclosed; (2) the number of deals that occurred within the same countries (in other words, deals in which acquirers and target companies are in the same countries); and (3) cross-border deals, which are deals between two different countries. These analyses help to determine the role of Western countries, in particular the U.S., and non-Western countries in the global capital market of the film industries.
5. Several media and telecommunications industries have been severely hit by the two consecutive economic crises and have never come back to the same degree of horizontal integration that occurred in 2000. For example, in the newspaper industry, there were 130 deals in 2000, a number that decreased to 84 in 2001, and there were still only 80 cases in 2009. In the broadcasting sector, 770 horizontal integrations occurred in 2000; however, there were only 571 deals in 2009. The Internet and software industries have decreased by as much as 61% during the same period, from 8,037 cases in 2000 to 3,132 deals in 2009. The result proves that the movie industries are the only sector shown to have rapidly recovered in the global M&As market. While other media and telecommunications industries have suffered from the economic crises, the movie industries had a transitional setback, followed by a swift recovery.
6. Five studios, 'The Big Five' (Warner Brothers, Paramount, 20th Century Fox, Loew's [MGM], and RKO [Radio-Keith-Orpheum]) worked to achieve vertical integration through the late 1940s, owning vast real estate on which to construct elaborate sets (Terra Media, 2010). However, the 1940s saw the system undermined by governmental trust-busting, televisualization, and suburbanization: The state called on Hollywood to divest ownership of theaters, even as the spread of television and housing from city centers diminished box office receipts. As such, vertical integration through ownership of production, distribution, and exhibition was outlawed domestically, but not on a global scale (Miller & Maxwell, 2006, pp. 36–37).

NOTES TO CHAPTER 7

This work was supported by the National Research Foundation of Korea Grant funded by the Korean Government (NRF-2010-332-B00648).

1. There are three most frequently used restructuring strategies: the spin-off, the split-off, and the sell-off. Spin-off is a type of divestiture, which is the

creation of an independent company through the sale or distribution of new shares of an existing business division of a parent company. This is where one firm divides itself into two or more firms and distributes ownership of the disintegrated firm to shareholders as a special dividend (Alexander & Owers, 2009). Meanwhile, split-off is a type of corporate reorganization whereby the stock of a subsidiary is exchanged for shares in a parent company. The sell-off is a relatively simple form of restructuring wherein a business unit within a media corporation is sold to another firm for cash or other considerations. It represents an acquisition for the purchasing firm. The disintegration model, which is another form of corporate transformation, has appeared in these various ways, but in many cases, they are not mutually exclusive.

2. Detailed reasons for the failure of convergence in terms of technological convergence between television and the Internet can be read in Castells (2000, pp. 189–190).
3. One of the major reasons for the unfinished deal was that the board of Walt Disney had rejected the $54 billion unsolicited stock offer from Comcast Corp. because Disney believed it was too low. However, Disney also certainly acknowledged that big media deals such as Vivendi-Universal and AOL-Time Warner had had their problems. They knew "on paper, marring content with distribution makes a lot of sense but in reality it's been a tough nut to crack" (Holson & Sorkin, 2004).

NOTES TO CHAPTER 8

1. George Cope originally joined BCE as president and CEO of Bell Canada, and Cope was named president and CEO of BCE Inc. and Bell Canada in July 2008.

References

Adams, E. (1995). Chain growth and merger waves: A macroeconomic historical perspective on press consolidation. *Journalism and Mass Communication Quarterly* 72(2): 376–389.
Adams, M. (2002, March). Making a merger work. *HR Magazines* 47(3): 52–57.
Advertising Age. (2002). Top 100 global marketers. 11 November, 30.
——— (2003). World's top 25 ad organizations. 21 April, 4.
——— (2012). Samsung launches biggest U.S. campaign to date for Galaxy S III. 20 June.
Adweek. (2009). Facebook to surpass MySpace in ad revenue. 23 December, 2.
Ahrens, F. (2004). GE, Vivendi give rise to a giant: New NBC Universal a $43 billion concern. *Washington Post*, 13 May, E1.
——— (2007). Newspaper-TV marriage shows signs of strain. *Washington Post*, 11 January.
——— (2008). Debt-saddled Tribune Co. files for bankruptcy protection. *Washington Post*, 9 December, D1.
Aksoy, A., and K. Robins. (1992). Hollywood for the 21st century: Global competition for critical mass in image markets. *Cambridge Journal of Economics 16*: 1–22.
Albarran, A. B., and R. K. Gormly. (2004). Strategic response or strategic blunder? An examination of AOL Time Warner and Vivendi Universal. In Picard, R. G. (Ed.), *Strategic Responses to Media Market Changes* (pp. 35–45). JIBS Research Reports Series, no. 2004–2.
Alexander, A., and J. Owers. (2009). Divestiture restructuring in the media industries: A financial market case analysis. *International Journal on Media Management 11*: 102–114.
Allbusiness.com. (2001). Vertical, horizontal integration change the face of television. Retrieved from http://www.allbusiness.com/technology/824024-1.html#ixzz1mPV5ligl
American Film Institute. (2010). Hollywood's Golden Age: An overview. Retrieved from http://www.fathom.com/course/10701053/session1.html
Anthony, C. (2009). AOL spin-off evokes spirit of IT bubble past. *Business Day*. 19 November.
Associated Press. (2009a). Time Warner walking out on AOL marriage. 28 May.
——— (2009b). Time Warner to spin off AOL unit. 29 May.
AT&T. (2002). *Annual Report 2002*. Bedminster, NJ: AT&T.
Bacchiocchi, E., M. Florio, and M. Gambaro. (2011). Telecom reforms in the EU: Prices and consumers' satisfaction. *Telecommunications Policy 35*: 382–396.
Bagdikian, B. (2002). *The New Media Monopoly*. Boston, MA: Beacon Press.
——— (2004). *The New Media Monopoly* (6th ed.). Boston, MA: Beacon Press.

Bakker, G. (2005). The decline and fall of the European film industry: Sunk costs, market size, and market structure, 1890–1927. *Economic History Review 58*(2): 310–351.

Baldwin, T. F., D. S. McVoy, & C. Steinfield. (1996). *Convergence: Integrating Media, Information, and Communication*. Thousand Oaks, CA: Sage.

The Banker. (2003). Africa slowdown in sell-offs. 1 April.

Bar, F., and C. Sandvig. (2008). US communication policy after convergence. *Media, Culture and Society 30*(4): 531–550.

Baran, P., and P. M. Sweezy. (1968). *Monopoly Capital: An Essay on the American Economic and Social Order.* New York: Monthly Review Press.

BBC News. (2000). Vivendi buys Seagram. 20 June. Retrieved from http://news.bbc.co.uk/2/hi/business/797146.stm

——— (2008). Nigeria phone privatization off. 18 February. Retrieved from http://news.bbc.co.uk/2/hi/business/7251154.stm .

Bell Canada Enterprises. (2005). Woodbridge and BCE announce new ownership structure for Bell Globemedia. Press Release. 2 December.

——— (2008). 2007 Annual Report. Montreal: Ball Canada Enterprise.

Belo. (2008). Belo Corp. completes spin-off of newspaper businesses. Press Release. February 5.

Belson, K. (2005). Justice department approves two big telecom deals. *New York Times*, 28 October, C4.

Blackstone, E., and G. Bowman. (1999). Vertical integration in motion pictures. *Journal of Communication 49*(1): 123–139.

Blair, H. (2001). You're only as good as your last job: The labor process and labor market in the British film industry. *Work, Employment and Society 15*(1): 149–169.

Bonney, B. (1984). Transnational advertising agencies and the local industry. *Media International Australia 31*: 34–39.

Bradbury, S., and K. Kasler. (November 2000). Verizon Communications: The merger of Bell Atlantic and GTE. *Corporate Finance*: 47.

Bradley, J. (2011). Agency 2011 report. *Advertising Age 82*(17): 24–41.

Business Week. (1996). The Global 1000. July, 133.

Cabanda, E., and M. Afiff. (2002). Performance gains through privatization and competition of Asian telecommunications. *ASEAN Economic Bulletin 19*(3): 255–256.

Cable & Wireless Communications. (2010). Demerger. Retrieved from http://www.cwc.com/past-present/demerger.html

Carabrese, A. (2008). *Privatization of the Media. The International Encyclopedia of Communication*. Malden, MA: Blackwell.

Carter, B. (2004). The media business: Deal complete, NBC is planning to cross-market. *New York Times*, 13 May. Retrieved from http://www.nytimes.com/2004/05/13/business/the-media-business-deal-complete-nbc-is-planning-to-cross-market.html

Castells, M. (2000). *The Rise of the Network Society*. Malden, MA: Blackwell.

——— (2001). *The Internet Galaxy*. New York: Oxford University Press.

CBC News. (2002). Teleglobe cuts 850 jobs, granted bankruptcy protection. Retrieved from http://www.cbc.ca/money/story/2002/05/15/teleglobe_bankruptcy_020515.html

Chakravartty, P., and D. Schiller. (2010). Neoliberal newspeak and digital capitalism in crisis. *International Journal of Communication 4*: 670–682.

Chalaby, J. (2008). Advertising in the Global Age: Transnational campaigns and pan-European television channels. *Global Media and Communication 4*(2): 139–156.

Chambers, T., and H. Howard. (2005). The economics of media consolidation. In A. B. Albarran, S. M. Chan-Olmsted, and M. Wirth (Eds.), *Handbook of Media Management and Economics* (pp. 363–386). Mahwah, NJ: Lawrence Erlbaum.

Chan-Olmsted, S. (1998). Mergers, acquisitions, and convergence: The strategic alliance of broadcasting, cable television, and telephone services. *Journal of Media Economics 11*(3): 33–46.

Chavez, R., M. Leiter, and T. Kiely. (2000). Should you spin off your Internet business? *Business Strategy Review 11*(2): 19–31.

Chintala, M. (2008). Cross-media companies are on the rise. Retrieved from http://www.indianmba.com/Faculty_Column/FC711/fc711.html

Cho, E. K. (2002). *A Comparative Study of the Business Model and Market Strategy of Global Media Group*. Seoul: Korean Broadcasting Institute.

Chon, B. S., J. H. Choi, G. A. Barnett, J. A. Danowski, and S. H. Joo. (2003). A structural analysis of media convergence: Cross-industry mergers and acquisitions in the information industries. *Journal of Media Economics 16*(3): 141–157.

Chozick, A. (2011). Times Co. negotiating to sell regional newspapers. *New York Times*, 19 December. Retrieved from http://mediadecoder.blogs.nytimes.com/2011/12/19/times-said-to-sell-regional-newspapers/

Chung, P. (2007). Hollywood domination of the Chinese kung fu market. *Inter-Asia Cultural Studies 8*(3): 414–424.

CNN/Money. (2004). Comcast bids for Disney. CNN news scripts. 18 February. Retrieved from CNN news scripts.

Coe, N., and J. Johns. (2004). Beyond production clusters: Towards a critical political economy of networks in the film and television industries. In D. Power and A. Scott (Eds.), *The Cultural Industries and the Production of Culture* (pp. 184–204). London: Routledge.

Collette, L., and B. Litman. (1997). The peculiar economies of new broadcast network entry: The case of United Paramount and Warner Bros. *Journal of Media Economics 10*(4): 3–22.

Dagmal, K. (2011). 100 Global Marketers. *Advertising Age 82*(43): 6–7.

Davis, A. (2011). Mediation, financialization, and the global financial crisis: An inverted political economy perspective. In D. Winseck and D. Y. Jin (Eds.), *The Political Economies of Media: Transformation of the Global Media Industries* (pp. 241–254). London: Bloomsbury.

De la Merced, M. (2009). With deal in hand, Charter files for bankruptcy. *New York Times*, 28 March, B3.

De Mooij, M. (2003). Convergence and divergence in consumer behavior: Implications for global advertising. *International Journal of Advertising 22*: 183–202.

Dennis, E. (2003). Prospects for a big idea—is there a future for convergence? *International Journal of Media Management 5*(1): 7–11.

——— (2006). Television's convergence conundrum: Finding the right digital strategy. *Television Quarterly 37*(1): 22–26.

Deuze, M. (2007). *Media Work*. Malden, MA.

De Vany, A., and H. McMillan. (2004). Was the antitrust action that broke up the movie studios good for the movies? Evidence from the stock market. *American Law and Economics Review, 6*(1): 135–153.

Ducoffe, R., and S. Smith. (1994). Mergers and acquisitions and the structure of the advertising agency industry. *Journal of Current Issues and Research in Advertising 16*(1): 15–26.

Dyer, G. (1984). Advertising-system and texts. *Media International Australia 31*: 5–12.

Dyer-Witheford, N., and Z. Sharman. (2005). The political economy of Canada's video and computer game industry. *Canadian Journal of Communication 30*: 187–210.

Edge, M. (2010). De-convergence and re-convergence in Canadian media. *The Convergence Newsletter* 7(10). Retrieved from http://www.sc.edu/cmcis/news/convergence/v7no10.html

eMarketer. (2010). Advertisers to spend $1.7 billion on social networks in 2010. Press Release. 16 August.

Endicott, R. (2003). 2003 *Advertising Age* agency report. *Advertising Age*, 21 April, 3.

Entman, R. (2010). Improving newspapers' economic prospects by augmenting their contributions to democracy. *International Journal of Press/Politics 15*(1): 104–125.

Epstein, G. (Ed.). (2005). *Financialization and the Global Economy*. Northampton, MA: Edward Elgar Press.

Erdal, I. J. (2007). Researching media convergence and crossmedia news production. *Nordicom Review 28*(2): 51–61.

Erturk, I., J. Froud, S. Johal, A. Leaver, and K. Williams. (2008). *Financialization at Work: Key Texts and Commentary*. London: Routledge.

Evers, J. (2000). AT&T to spin off Liberty Media. *Networkworld*. 16 November. Retrieved from http://www.networkworld.com/news/2000/1116attspin.html

Faulkner, R., and A. Anderson. (1987). Short-term projects and emergent careers: Evidence from Hollywood. *American Journal of Sociology* 92(4): 879–909.

Financial Times. (1998). Bertelsmann purchase set to open a whole net chapter. 25 March, 18.

Fine, J., and B. Johnson. (2005). The big-bang: Viacom's proposed split raises the question: Has consolidation failed to deliver on its promise, and could it all come undone? *Advertising Age*, 21 March, 33.

Flew, T. (2007). *Understanding Global Media*. New York: Palgrave.

Foster, J. B. (2007). The financialization of capitalism. *Monthly Review 58*(11): 1–12.

Frean, A. (2010). The Miramax brothers tell Disney: Let us buy it back. *The Times*, 10 April, 33.

Frieden, R. (2001). *Managing Internet-Driven Change in International Telecommunications*. Boston: Artech House.

Friedman, M. (1982). *Capitalism and Freedom*. Chicago: University of Chicago Press.

Friedman, M. (2002, special edition). *Capitalism and Freedom*. Chicago: University of Chicago Press.

Fu, W. (2009). Screen survival of movies at competitive theaters: Vertical and horizontal integration in a spatially differentiated market. *Journal of Media Economics* 22(2): 59–80.

Fung, A. (2006). Think globally, act locally: China's rendezvous with MTV. *Global Media and Communication* 2(1): 71–88.

García Avilés, J. A., and M. Carvajal. (2008). Integrated and cross-media newsroom convergence: Two models of multimedia news production—the cases of Novotécnica and La Verdad Multimedia in Spain. *Convergence 14*(2): 221–239.

Garnham, N. (1990). *Capitalism and Communication: Global Culture and the Economics of Information*. London: Sage.

——— (2011). The political economy of communication revisited. In J. Wasko, G. Murdock, and H. Sousa (Eds.). *The Handbook of Political Economy of Communications* (pp. 41–61). Malden, MA: Wiley-Blackwell.

George, E. (2010). Reading the notion of convergence in light of recent changes to the culture and communication industries in Canada. *Canadian Journal of Communication 35*: 556–564.

Geradin, D., and D. Luff. (2004). Introduction. In D. Geradin and D. Luff (Eds.), *The WTO and Global Convergence in Telecommunications and Audio-Visual Service* (pp. 3–20). Cambridge: Cambridge University Press.

Gershon, R. (2006). Deregulation, privatization and the changing global media environment. In G. Murdoch and J. Wasko (Eds.), *Media in the Age of Marketization* (pp. 185–205). Cresskill, NJ: Hampton Press.

Gil, A. (2008). Breaking the studios: Antitrust and the motion picture industry. *NYU Journal of Law and Liberty 3*(83): 83–123.

Global ad-spend trends: World advertising expenditure. (2001). *International Journal of Advertising 20*(2): 266–268.

Global Insights Inc. (2002). U.S. advertising market predicted to grow 6% in 2003. Retrieved from http://www.globalinsight.com/About/PressRelease/PressRelease209.htm

Global Telecoms Business. (2010). Turkey may sell Turk Telecom stake. 4 January.

——— (2011). Telecom NZ shareholders OK split. 1 October.

Goldman, C., I. Gotts, and M. Piaskoski. (2003). The role of efficiencies in telecommunications merger review. *Federal Communications Law Journal 56*: 86–151.

Gomery, D. (1986). Vertical integration, horizontal regulation: The growth of Rupert Murdoch's US media empire. *Screen 27*(3/4): 78–86.

Gomery, D. (2000). The Hollywood film industry: Theatrical exhibition, pay TV, and home video. In B. Campaign and D. Gomery (Eds.), *Who Owns the Media: Competition and Concentration in the Mass Media Industry* (pp. 359–436). Mahwah, NJ: Lawrence.

Google. (2012). Facts about Google's acquisition of Motorola. Retrieved from http://www.google.com/press/motorola/

Gordon, R. (2003). The meanings and implications of convergence. In K. Kawamoto (Ed.), *Digital Journalism: Emerging Media and the Changing Horizons of Journalism*. Lanham, MD: Rowman & Littlefield.

Grant, P. S., and C. Wood. (2004). *Blockbusters and Trade Wars: Popular Culture in a Globalized World*. Toronto: Douglas & Mclnytre.

Guback, T. (1987). The evolution of the motion picture theater business in the 1980s. *Journal of Communication 37*(2): 60–77.

Habermas, J. (1989). *The Structural Transformation of the Public Sphere: An Inquiry Into a Category of Bourgeois Society*. (Trans. Thomas Burger with the assistance of Frederick Lawrence). Cambridge, MA: MIT Press.

Hall, H. (1996). The West and the rest. In Stuart Hall, David Held and Kenneth Thompson (Eds.), *Modernity: An Introduction to Modern Societies* (pp. 185–227). Malden, MA: Blackwell.

Harmetz, A. (1986). Hollywood seeks control of outlets. *New York Times*. 3 March, D1.

Hart-Landsberg, M. (2006). Neoliberalism: Myths and reality. *Monthly Review 57*(11): 1–17.

Hazelton, J. (2000). China opens doors to Hollywood. *Screen International* (1261), 2 June, 8.

Herman, E., and R. McChesney. (1997). *The Global Media: The New Missionaries of Corporate Capitalism*. New York: Cassell.

Herrera-Gonzalez, F., and L. Martin. (2009). The endless need for regulation in telecommunication: An explanation. *Telecommunications Policy 33*: 664–675.

Holson, L., and A. R. Sorkin. (2004). Disney board rejects bid from Comcast as too low. *New York Times*, 17 February. Retrieved from http://www.nytimes.com/2004/02/17/business/disney-board-rejects-bid-from-comcast-as-too-low.html

Hope, W. (2011). Global capitalism, temporality, and the political economy of communication. In J. Wasko, G. Murdock, and H. Sousa (Eds.), *The Handbook of Political Economy of Communications* (pp. 521–540). Malden, MA: Wiley-Blackwell.

Horwitz, R. (2005). On media concentration and the diversity question. *Information Society 21*: 181–204.

Howard, H. (1998). The 1996 Telecommunications Act and TV station ownership: 1 year later. *Journal of Media Economics 11*(3): 21–32.

Huang, J. S., and D. Heider. (2008). Media convergence: A case study of a cable news station. *International Journal on Media Management* 9(3): 105–115.
IHS iSuppli (2012). Smartphones see accelerated rise to dominance. Press Release. 28 August.
Ikenberry, G. (2007). Globalization as American hegemony. In D. Held and A. McGrew (Eds.), *Globalization Theory: Approaches and Controversies* (pp. 41–61). Malden, MA: Polity Press.
Indiaserver. (2009). Motion Picture Association introduces its first office in India. Retrieved from http://www.india-server.com/news/motion-picture-association-introduces-6194.html
International Telecommunication Union. (2002). *World Telecommunication Development Report*. Geneva: ITU.
——— (2003a). *Cellular Subscribers*. Geneva: ITU.
——— (2003b). *Main Telephone Lines*. Geneva: ITU.
——— (2010). *World Telecommunication/ICT Development Report 2010*. Geneva: ITU.
——— (2011). *World Telecommunication/ICT Development Report 2011*. Geneva: ITU.
——— (2012). *Key Global Telecom Indicators for the World Telecom Service Sector*. Geneva: ITU. Retrieved from http://www.itu.int/ITU-D/ict/statistics/at_glance/KeyTelecom.html
Internet Business News. (2009). Time Warner completes AOL spin-off. 10 December. Retrieved from http://www.informationweek.com/internet/ebusiness/aol-completes-spin-off-from-time-warner/222001597
Iosifidis, D. (2005). Digital switchover and the role of BBC services in digital TV take-up. *Convergence: International Journal of Research Into New Media Technologies* 11(3): 57–74.
Jenkins, H. (2001). Convergence? I Diverge. *Technology Review 93*. Retrieved from http://www.technologyreview.com/business/12434/
——— (2006). *Convergence Culture*. New York: New York University Press.
Jihong, W., and R. Kraus. (2002). Hollywood and China as adversaries and allies. *Pacific Affairs 75*(3): 419–434.
Jin, D. Y. (2005). The telecom crisis and beyond: Restructuring of the global telecommunications system. *International Communication Gazette 67*(3): 289–304.
——— (2006a). Cultural politics in Korea's contemporary films under neoliberal globalization. *Media, Culture & Society 28*(1): 6–23.
——— (2006b). Political and economic processes in the privatization of the Korean telecommunications industry: A case study of Korea telecom, 1987–2003. *Telecommunications Policy 30*(1): 3–13.
——— (2007). Transformation of the world television system under neoliberal globalization, 1983–2003. *Television and New Media 8*(3): 179–196.
——— (2008). Neoliberal restructuring of the global communication system: Mergers and acquisitions. *Media, Culture & Society 30*(3): 357–373.
——— (2010). Financialization of the Asian game industry in the midst of the global recession. *Iowa Journal of Communication 42*(1): 23–44.
——— (2011a). De-convergence and deconsolidation in the global media industries: The rise and fall of (some) media conglomerates. In D. Winseck and D. Y. Jin (Eds.), *The Political Economies of Media: Transformation of the Global Media Industries* (pp. 167–182). London: Bloomsbury.
——— (2011b). *Hands On/Hands Off: The Korean State and the Market Liberalization of the Communication Industry*. Cresskill, NJ: Hampton Press.
——— (forthcoming). Critical analysis of user commodities as free labor in social networking sites: A case study of cyworld. *Continuum: Journal of Media and Cultural Studies*.

Jonquireres, G. D. (1997). Template for trade tariffs. *Financial Times*, 18 February, 6.
Jubak, J. (2002, 16 October). Mergers and acquisitions made in the 1990s because of supposed synergies are beginning to fall apart in this decade. CNBC Show: Business Center. CNBC News Transcripts.
Jung, J. M. (2004). Acquisitions or joint ventures: Foreign market entry strategy of U.S. advertising agencies. *Journal of Media Economics 17*(1): 35–50.
Kamitschnig, M. (2006). After years of pushing synergy, Time Warner Inc. says enough. *Wall Street Journal*, 2 June, A1.
Kawashima, N. (2009). The structure of the advertising industry in Japan: The future of the mega-agencies. *Media International Australia 133*: 75–84.
Keane, M. (2006). From made in China to created in China. *International Journal of Cultural Studies 9*(3): 285–296.
Kelsey, J. (2006). Globalization of cultural policymaking and the hazards of legal seduction. In G. Murdoch and J. Wasko (Eds.), *Media in the Age of Marketization* (pp. 151–181). Cresskill, NJ: Hampton Press.
Kirchhoff, S. (2010). The U.S. newspaper industry in transition. Congressional Research Service. 9 September.
Klein, C. (2003). The Asian factor in Hollywood: Breaking down the notion of a distinctly American cinema. Literature in MIT. Retrieved from http://web.mit.edu/lit/www/spotlightarticles/ hollywood.html
Knight, B. (2010). Movie industry debate says Hollywood isn't going anywhere. Retrieved from http://www.dw-world.de/dw/article/0,,5267505,00.html
Ko, K. M., and H. Y. Cha. (2009). The globalization of the Korean advertising industry: Dependency or hybridity. *Media International Australia 133*: 93–109.
Kolo, C., and P. Vogt. (2004). Traditional media and their Internet spin-offs: An explorative study of key levers for online success and the impact of offline reach. *International Journal on Media Management 6*(1&2): 23–35.
Korea Communications Commission. (2012). Telecommunications data. Seoul: KCC.
Korea Creative Contents Agency. (2010). *2010 Korean Game Whitepaper.* Seoul: Korea Creative Contents Agency.
——— (2012). *2011 Korean Game Whitepaper.* Seoul: Korea Creative Contents Agency. *The Korea Herald*. (2000). Seoul to increase foreign ownership ceiling for KT. 22 September.
Korean Film Council. (2010). *2009 Korean Film Yearbook*. Seoul: Korean Film Council.
Korea Times. (1998). Foreigners allowed to invest in media. 21 December, 2.
Kouwe, Z. (2008). Takeover bid for Bell Canada evaporates. *New York Times*, 10 December. Retrieved from http://www.nytimes.com/2008/12/11/business/worldbusiness/11place.html
Krantz, M. (2005). Many spinoffs prove popular with investors. *USA Today*, 16 June. Retrieved from http://usatoday30.usatoday.com/money/markets/us/2005-06-15-spinoffs-usat_x.htm
——— (2008). Time Warner adds to spinoff trend with plans for cable unit; Shareholders to get ownership. *USA Today*, 22 May, 4B.
Krippner, G. (2005). The financialization of the American economy. *Socio-Economic Review 3*(2): 173–208.
Kutz, W. (2007). *Culture Conglomerates: Consolidation in the Motion Picture and Television Industries*. Lanham, MD: Rowman & Littlefield.
Latour A., and S. Young. (2005). Boards of SBC and AT&T approve $16 billion deal. *Wall Street Journal.* 31 January. Retrieved from http://online.wsj.com/article/0,,SB110711095360640355.html
Lawson-Borders, G. (2003). Integrating new media and old media: Seven observations of convergence as a strategy for best practices in media organization. *International Journal of Media Management 5*(2): 91–99.

Lazonick, W. (2009). *Sustainable Prosperity in the New Economy*. Kalamazoo, MI: W. E. Upjohn Institute for Employment Research.
Lee, B. J. (2002). WPP nears accord for S. Korea's LG AD. *Advertising Age 73*(39): 10–11.
Lee, J. Y. (2007). Theaters go to multinational competitions. *Sisa Journal*, 30 July.
Lee, M. (2012). Nexon buys $685 million stake in NCSoft, becomes biggest holder. *Bloomberg*, 8 June. Retrieved from http://www.bloomberg.com/news/2012-06-08/nexon-buys-685-million-stake-in-ncsoft-becomes-biggest-holder.html
Lee, S. B. (1998). The political economy of the Russian newspaper industry. *Journal of Media Economics 11*(2): 51–71.
Li, F., and J. Whalley. (2002). Deconstruction of the telecommunications industry: From value chains to value networks. *Telecommunications Policy* 26: 451–472.
Liao, W. (2009, 10 March). More than 10 newspaper groups plan to float this year. *National Business Daily*. Retrieved July 20, 2010, from http://news.ccidnet.com/art/952/20090310/1702989_1.html
Lieberman, D. (2006). McClatchy to buy Knight Ridder for $4.5 billion. *USA Today*, 13 March. Retrieved from http://usatoday30.usatoday.com/money/media/2006-03-13-knight-ridder_x.htm
Lindio-McGovern, L. (2007). Neo-liberal globalization in the Philippines: Its impact on Filipino women and their forms of resistance. *Journal of Developing Societies 23*(1/2): 15–35.
Liu, F., and S. Chan-Olmsted. (2002). Partnerships between the old and the new: Examining the strategic alliances between broadcast television networks and Internet firms in the context of convergence. *International Journal on Media Management 5*(1): 47–56.
Lorenzen, M. (2008). On the globalization of the film industry. Creative Encounters Working Paper #8. Copenhagen Business School.
Machlup, F. (1962). *The Production and Distribution of Knowledge in the United States*. Princeton, NJ: Princeton University Press.
Macmillian, R. (2007). Belo to spin off newspaper business, shares surge. *Reuters*. 1 October. Retrieved from http://www.reuters.com/article/2007/10/01/us-belo-spinoff-idUSWNAS521120071001
Magder, T. (2004). Transnational media, international trade and the idea of cultural diversity. *Continuum: Journal of Media & Cultural Studies 18*(3): 380–397.
Mahamdi, Y. (1992). Television, globalization, and cultural hegemony: The evolution and structure of international television. Dissertation. The University of Texas at Austin.
Maich, S. (2005). Better off without you. *Maclean's 118*(20), 16 May, 33.
Mallikarjunappa, T., and P. Nayak. (2007). Why do mergers and acquisitions quite often fail? *AIMS International Journal of Management 1*(1): 53–69.
Malone, M. (2010). Station M&A: Wait till next year. *Broadcasting & Cable 140*(15): 22.
Maney, K. (1997, 20 March). America is building British phone empire. *USA Today*, 4B.
McChesney, R. (1999). *Rich Media, Poor Democracy: Communication Politics in Dubious Times*. Urbana, IL: University of Illinois Press.
——— (2001). Global media, neoliberalism, and imperialism. *Monthly Review 52*(19): 1–19.
——— (2008). *The Political Economy of Media: Enduring Issues, Emerging Dilemmas*. New York: Monthly Review Press.
McClintock, P. (2005). Clear Channel doing the splits. *Daily Variety*, 2 May, 1.
McDonnell, J., and J. Silver. (2009). Hollywood dominance: Will it continue? Paper presented at What Is Film? Change and Continuity in the 21st Century, 6–7 November, Portland, Oregon.

McLarty, T. (1998). Liberalized telecommunications trade in the WTO: Implications for universal service policy. *Federal Communications Law Journal 5*(1): 1–59.
Mcnish, J. (2008). BCE unveils 100-day plan as Cope takes the helm. *The Globe and Mail*, 11 July.
McPhail, T. (2006). *Global Communication: Theories, Stakeholders, and Trends* (2nd ed.). Oxford: Blackwell.
Melinda, R. (2011). New technologies: Convergence, the new way of doing journalism. *Journalism si communicare 11*(1): 1–7.
Miller, T., N. Govil, J. McMurria, R. Maxwell, and T. Wang. (2005). *Global Hollywood 2*. London: BFI Publishing.
Miller, T., and R. Maxwell. (2006). Film and globalization. In O. Boyd-Barrett (Ed.), *Communications Media Globalization and Empire* (pp. 33–52). New Barnet: John Libbey.
Mosco, V. (2004). *The Digital Sublime: Myth, Power and Cyberspace*. Cambridge: MIT Press.
——— (2005). Here today, outsourced tomorrow: Knowledge workers in the global economy. *Javnost-the public 12*(2): 39–56.
——— (2008). Current trends in the political economy of communication. *Global Media Journal Canadian Edition 1*(1): 45–63.
——— (2009). *The Political Economy of Communication* (2nd ed.). London: Sage.
Mosco, V., and C. McKercher. (2006). Convergence bites back: Labour struggles in the Canadian communication industry. *Canadian Journal of Communication 31*: 773–751.
MPAA. (1986). General theatrical data. Los Angeles, CA: MPAA.
——— (2009). The economic impact of the motion picture and television industry on the U.S. Los Angeles, CA: MPAA.
——— (2011). 2010 theatrical market statistics. Los Angeles, CA: MPAA.
Muehlfeld, K., P. Sahib, and A. Witteloostuijin. (2007). Completion or abandonment of mergers and acquisitions: evidence from the newspaper industry, 1981–2000. *Journal of Media Economics 20*(2): 107–137.
Mueller, M. (1999). Digital convergence and its consequences. *Javnost-the public 6*(3): 11–28.
Mueller, M. (2004). Convergence: A reality check. In D. Geradin and D. Luff (Eds.), *The WTO and Global Convergence in Telecommunications and Audio-Visual Service* (pp. 311–322). Cambridge: Cambridge University Press.
Mulligan, C. (2012). *The Communications Industries in the Era of Convergence*. London: Routledge.
Murdock, G. (2006). Notes from the number one country: Herbert Schiller on culture, commerce, and American power. *International Journal of Cultural Policy 12*(2): 209–227.
——— (2011). Political economies as moral economies: Commodities, gifts, and public goods. In J. Wasko, G. Murdock, and H. Sousa (Eds.). *The Handbook of Political Economy of Communications* (pp. 13–40). Malden, MA: Wiley-Blackwell.
Murray, S. (2003). Media convergence's third wave. *Convergence 9*(1): 8–18.
Musgrove, M. (2009). Time Warner's talk of AOL spin-off stirs hope for Internet unit's future. *Washington Post*. 30 April, A16.
National Association of Theater Owners. (2010). Number of U.S. movie screens. Retrieved from http://www.natoonline.org/statisticsscreens.htm
Nerone, J. (2011). Does journalism history matter? *American Journalism 28*(4): 7–27.
NewsHour. (2000). Tribune buys Times Mirror. 21 March. Retrieved from http://www.pbs.org/newshour/bb/media/jan-june00/tribune_3–21.html

Ng, J. (2011). Global telecom revenues hit $1.8 T. *ZDNet*, 10 May. Retrieved from http://www.zdnetasia.com/global-telecom-revenues-hit-1–8t_print-62300288.htm
Nielsen. (2012). Global report: Multi-screen media usage. Press Release. 15 May.
Noam, E. (2006). How to measure media concentration. FT.com, 19 June. Retrieved from http://www.ft.com/cms/s/2/da30bf5e-fa9d-11d8-9a71-00000e2511c8.html#axzz2BmBxtIaJ
Noam, E. (2009). *Media Ownership and Concentration in America*. New York: Oxford University Press.
Noble, C. (2003). Restructuring debts Vivendi's revenue. *Toronto Star*, 8 November, C4.
Noll, M. (2003). The myth of convergence. *International Journal of Media Management 5*(1): 12–13.
NORDICOM. (2010). *A Sampler of International Media and Communication Statistics 2010*. Gothenburg, Sweden: Nordicom.
Oba, G., and S. Chan-Olmsted. (2006). Self-dealing or market transaction? An exploratory study of vertical integration in the U.S. television syndication market. *Journal of Media Economics 19*(2): 99–118.
O'Brien, K. (2007). Breakup of Eircom: An opening for rivals? *International Herald Tribune*, 24 December, 10.
Oh, J. G. (1996). Global strategic alliances in the telecommunications industry. *Telecommunications Policy 20*(9): 713–720.
Oram, R., and N. Tait. (1989). Ogilvy agrees dollar 865 M WPP bid. *Financial Times*, 17 May, 1.
Organisation for Economic Co-operation and Development. (1990). Communication outlook 1990. Paris: OECD.
——— (1997). Communication outlook 1997. Paris: OECD.
——— (1999). Communication outlook 1999. Paris: OECD.
——— (2001). Communication outlook 2001. Paris: OECD.
——— (2005). OECD Communication outlook 2005. Paris: OECD.
——— (2007). Communication outlook 2007. Paris: OECD.
——— (2010a). News in the Internet Age: New trends in news publishing. Paris: OECD.
——— (2010b). The evolution of news and the Internet. Paris: OECD.
——— (2011). Communication outlook 2011. Paris: OECD.
Peltier, S. (2004). Mergers and acquisitions in the media industries: Were failures really unforeseeable? *Journal of Media Economics 17*(4): 261–278.
Pfanner, E. (2009). Foreign films get a hand from Hollywood. *New York Times*, 17 May.
Picard, R. (2004). Commercialism and newspaper quality. *Newspaper Research Journal 25*(1): 54–65.
Pickering, J. F. (1983). The causes and consequence of abandoned mergers. *Journal of Industrial Economics 31*(3): 267–281.*l*
PLDT. (2011). Corporate governance. Retrieved from http://www.pldt.com.ph/governance/Pages/Subsidiaries.aspx
Prashad, S. (2006). Bell to convert to huge income trust. *Toronto Star*, 12 October, C01.
PricewaterhouseCoopers. (2000). *Global entertainment and media outlook 2000–2004*. New York: PricewaterhouseCoopers LLP.
——— (2009). *Global entertainment and media outlook 2009–2013*. New York: PricewaterhouseCoopers LLP.
Pritchard, T. (2000a). BCE of Canada offers $6.7 billion to buy the rest of Teleglobe. *New York Times*, 16 February, C16.
——— (2000b). Biggest Canadian phone company offers to buy TV network. *New York Times*, 26 February, C2.

Rainie, L., and B. Wellman. (2012). *Networked: The New Social Operating System.* Cambridge, MA: MIT Press.
Reed, B. (2008). Global telecom revenue to hit $2 trillion in 2008. *Network World*, 17 September. Retrieved from http://www.networkworld.com/news/2008/091708-global-telecom.html
Robins, A. (1993). Organization as strategy-restructuring production in the film industry. *Strategic Management Journal 14*: 103–118.
Rusli, E., and P. Eavis. (2012). Facebook raises $16 billion in I.P.O. *New York Times*. 17 May. Retrieved from http://dealbook.nytimes.com/2012/05/17/facebook-raises-16-billion-in-i-p-o/
Satoshi, I. (2007). A Western contention for Asia-centered communication scholarship paradigms: A commentary on Gordon's paper. *Journal of Multicultural Discourses*, 2(2): 108–114.
Scherer, E. (2010). Context is king. *AFP-MediaWatch* 7: 4–14. Retrieved from http://mediawatch.afp.com /public/AFP-MediaWatch_Automne-Hiver-2009–2010.pdf
Schiller, D. (1999a). Deep impact: The Web and the changing media economy. *Info: Journal of Policy, Regulation and Strategy for Telecommunications, Information and Media* 1(1): 35–51.
——— (1999b). *Digital Capitalism*. Cambridge, MA: MIT Press.
——— (2001). World communications in today's age of capital. *Emergences 11*(1): 51–68.
——— (2003). The telecom crisis. *Dissent 50*(1): 66–70.
——— (2007). *How to Understand Information*. Urbana, IL: University of Illinois Press.
——— (1989). *Culture Inc.: The Corporate Takeover of Public Expression*. New York: Oxford University Press.
Schnaars, S., G. Thomas, and C. Irmak. (2008, June). Predicting the emergence of innovations from technological convergence: Lessons from the twentieth century. *Journal of Macromarketing 28*: 157–168.
Shan, S. (2011). GTV has license renewal application approved with programming condition. *Taipei Times*, 29 December, 2.
Sherman, J. (2006). CBS Corp. in the fish bowl; Industry watching for how net evolves. *Television Week*, 9 January, 17.
Simpson, P. (2012). China bans foreign programmes from prime time television. *The Telegraph*, 14 February. Retrieved from http://www.telegraph.co.uk/news/worldnews/asia/china/9081577/China-bans-foreign-programmes-from-prime-time-television.html
Sinclair, J. (2008). Globalization and the advertising industry in China. *Chinese Journal of Communication 1*(1): 77–90.
Sklair, L. (2001). *The Transnational Capitalist Class*. Oxford: Blackwell.
Soderlund, W., and W. Romanow. (2005). Failed attempts at regulation of newspaper ownership. In W. Soderlund and K. Hildebrandt (Eds.), *Canadian Newspaper Ownership on the Era of Convergence: Rediscovering Social Responsibility* (pp. 11–30). Edmonton, Alberta: University of Alberta Press.
Soja, E. (1989). *Postmodern Geographies: The Reassertion of Space in Critical Social Theory*. London: Verso.
Sommerville, Q. (2000, 16 April). BT goes its own separate ways. *Scotland on Sunday*, 7.
Sorkin, A., and K. Belson. (2005). Alltel plans to spin off phone unit. *New York Times*, 9 December. Retrieved from http://www.nytimes.com/2005/12/09/business/09phone.html?gwh=595CEC3970D69DB9C62B40A1890979E4&_r=0
Standard & Poor's. (2002, April). A year of turmoil for U.S. long-distance carriers. *Industry Surveys*: 13.
——— (2003, April). Telecom industry enters new era. *Industry Surveys*: 3.

——— (2008, February 7). Telecommunications: Wireline. *Industry Surveys* Retrieved from http://homepage.smc.edu/thomas_phillip/rpt/TelecommunicationsWireline.pdf

Starr, P. (2002, 9 September). The great telecom implosion. *American Prospect 13*(16): 21–27.

Stelter, B. (2011). Ownership of TV sets falls in U.S. *New York Times*. 3 May. Retrieved from http://www.nytimes.com/2011/05/03/business/media/03television.html?gwh=6E898C4BA4C937CAD4BCBE43EACA67C3

Stevenson, R. W. (1985). Economic scene: A rush to own TV properties. *New York Times*, 18 December, D2.

Storch, C. (1985). Capital One's to take over ABC. *Chicago Tribune*, 19 March.

Storper, M., and S. Christopherson. (1987). Flexible specialization and regional industrial agglomeration: The US film industry. *Annals of the Associations of American Geographers 77*(1): 104–117.

Straubhaar, J. (2001). Brazil: The role of the state in world television. In N. Morris and S. Waisbord (Eds.), *Media and Globalization: Why the State Matters* (pp. 133–154). New York: Rowman & Littlefield.

Su, W. (2010). To be or not to be?—China's cultural policy and counter-hegemony strategy toward global Hollywood from 1994 to 2000. *Journal of International and Intercultural Communication 3*(1): 38–58.

Sunada, M. (2010).Vertical integration in the Japanese movie industry. *Journal of Industry, Competition, and Trade 10*(2): 135–150.

Teather, D. (2005). Analysis: Viacom signals the end of the road for media juggernauts. *The Guardian*, 18 March, 23.

Terra Media. (2010). *Vertical Integration in the Film Industry.* Available at www.terramedia.co.uk/media/film/vertical_integration.htm

Thierer, A. (2005). *Media Myths: Making Sense of the Debate Over Media Ownership*. Washington DC: Progress & Freedom Foundation.

Thornton, L., and S. Keith. (2009). From vonvergence to webvergence: Tracking the evolution of broadcast-print partnerships through the lens of change theory. *Journalism and Mass Communication Quarterly 86*(2): 257–276.

Thussu, D. (2006). *International Communication: Continuity and Change* (2nd ed.). New York: Hodder Arnold.

The Times. (2003). Time Warner drops AOL. 19 September, B27.

Time Warner Cable. (2008). Time Warner and Time Warner Cable agree to separation. Press Release. 21 May.

The Top 100 Deals of 1998. (1999, March/April). *Mergers & Acquisitions 33*(5): 33–37.

The Top 100 Deals of 1999. (2000), February). *Mergers & Acquisitions 35*(2): 21–24.

The Top 100 Deals of 2000. (2001, February). *Mergers & Acquisitions 36*(2): 18–21.

The Top 100 Deals of 2001. (2002, February). *Mergers & Acquisitions 37*(2): 14–17.

The Top 100 Deals of 2002. (2003, February). *Mergers & Acquisitions 38*(2): 16–19.

The Top 100 Deals of 2003. (2004, February). *Mergers & Acquisitions 39*(2): 10–13.

The Top 100 Deals of 2004. (2005, February). *Mergers & Acquisitions 40*(2): 30–33.

The Top 100 Deals of 2005. (2006, February). *Mergers & Acquisitions 41*(2): 32–35.

The Top 100 Deals of 2006. (2007, February). *Mergers & Acquisitions 42*(2): 46–50.

The Top 100 Deals of 2007. (2008, February). *Mergers & Acquisitions 43*(2): 62–66.

Trillas, F. (2002). Mergers, acquisitions and control of telecommunications firms in Europe. *Telecommunications Policy 26*(5/6): 260–286.

Tucker, E. (1985). Weighs role in hostile bids for TV networks; Decisions to shape industry's future. *Washington Post*, 21 July.

Uchitelle, L. (2002). Turmoil at WorldCom: The work force: Job cuts take heavy toll on telecom industry. *New York Times*, 29 June, C1.

Ulset, S. (2007). Restructuring diversified telecom operators. *Telecommunications Policy 31*: 209–229.
UNESCO. (1999). *UNESCO's Statistical Yearbook 1999*. Paris: UNESCO.
United Nations. (1979). *Transnational Corporations in Advertising: Technical Report.* New York: United Nations.
United Nations Statistical Office. (1972). *Statistical Yearbook*. New York: United Nations.
U.S. Department of Commerce. (2002). Survey of current business. Washington DC: Department of Commerce.
——— (2011). Survey of current business. Washington DC: Department of Commerce.
U.S. Senate Commerce, Science, and Transportation Committee. (2002). Panel of a hearing on financial turmoil in the telecom marketplace, 30 July, 16–17.
Van Dijk, J. (2006). *The Network Society* (2nd ed.). London: Sage.
Vascellaro, J. (2009). Yahoo plans to shutter GeoCities. *Wall Street Journal*, 24 April, B4.
Vega, T. (2011). Times Co. agrees to sell regional newspaper group. *New York Times*, 27 December. Retrieved from http://mediadecoder.blogs.nytimes.com/2011/12/27/times-company-sells-regional-newspaper-group/?gwh=3F6B3DEF89768FFF92E73A0520B19163
Venn, C., and M. Featherstone. (2006). Modernity. *Theory, Culture and Society, 23*(2/3): 457–476.
Wang, G. (2003). Foreign investment policies, sovereignty and growth. *Telecommunications Policy* 27(1): 267–282.
Wasko, J. (2003). *How Hollywood Works*. London: Sage.
Wasko, J., and M. Erickson (Eds.). (2008). *Cross-Border Cultural Production*. New York: Cambria Press.
Waterman, D. (2000). CBS-Viacom and the effects of media mergers: An economic perspective. *Federal Communication Law Journal 52*(3): 531–550.
Waterson, M. (1992). International advertising expenditure statistics. *International Journal of Advertising 11*(2): 14–67.
Watkins, M. (2008). Strange bedfellows at BCE: Ontario teachers and U.S. private equity funds. In M. Moll and L. R. Shade (Eds.), *For $ale to the Highest Bidder: Telecom Policy in Canada*. (pp. 56–60). Ottawa: Canadian Centre for Policy Alternatives.
Weber, J. (2008). The Bell system divestiture: Background, implementation, and outcome. *Federal Communications Law Journal 61*(1): 21–30.
Wee, G. (2007). Time to cut 2,000 jobs at AOL: Turns focus to ads. *Financial Post*, 16 October F16.
Wellenius, B., C. Braga, and C. Qiang. (2000). Investment and growth of the information infrastructure: Summary results of a global survey. *Telecommunications Policy 24*(8.9): 639–643.
Weston, J. F., M. L. Mitchell, and J. H. Mulherin. (2004). *Takeovers, Restructuring, and Corporate Governance* (4th ed.). Upper Saddle River, NJ: Prentice Hall.
Winseck, D. (2002).Netscapes of power: Convergence, consolidation and power in the Canadian mediascape. *Media, Culture and Society 24*(6): 795–819.
——— (2010). Financialization and the "crisis of the media": The rise and fall of (some) media conglomerates in Canada. *Canadian Journal of Communication 35*: 365–393.
——— (2011). Financialization and the crisis of the media: The rise and fall of (some) media conglomerates in Canada. In D. Winseck and D. Y. Jin (Eds.), *The Political Economies of Media: Transformation of the Global Media Industries* (pp. 142–166). London: Bloomsbury.

Winseck, D., and R. Pike. (2007). *Communication and Empire*. Durham, NC: Duke University Press.

Wirtz, B. W. (2001). Reconfiguration of value chains in converging media and communications markets. *Long Range Planning, 34*(4): 489–506.

World Association of Newspapers and News Publishers. (2010). World press trends 2010. Press Release. 5 August.

——— (2011). World press trends 2010. Paris: World Association of Newspapers and Publishers.

——— (2012). World Press Trends: Newspaper Audience Rise, Digital Revenues Yet to Follow. Press Release. 3 September.WPP. (2012). WPP at a glance. Retrieved from http://www.wpp.com/wpp/about/wppataglance/

Yang, C., and T. Lowry. (2001). AOL Time Warner: Aiming two high. *Business Week*, 3748, 10 September, 98–100.

Yoo, S. H. (2005). Korea a major player in Asia-Pacific ad market. *Korean Herald*, 17 March.

ZenithOptimedia. (2002). Signs of recovery. Press Release. 9 December.

——— (2011). Global ad expenditure. Press Release. 3 October.

——— (2012). Global advertising growth continues as Latin America and Asia Pacific compensate for weakening Europe. Press Release. 15 March.

Zhang, S. I. (2010). Chinese newspaper ownership, corporate strategies, and business models in a globalizing world. *International Journal on Media Management 12*: 205–230.

Zhao, Y. (1998). *Media, Market, and Democracy in China: Between the Party Line and the Bottom Line*. Urbana: University of Illinois Press.

——— (2008). Neoliberal strategies, socialist legacies: communication and state transformation in China. In P. Chakravartty and Y. Zhao (Eds.), *Global Communications: Towards a Transcultural Political Economy* (pp. 25–30). New York: Rowman & Littlefield.

Index

V

W

X

Y

Z